SIMULATED VOYAGES

Using Simulation Technology to
Train and License Mariners

Committee on Ship-Bridge Simulation Training

Marine Board

Commission on Engineering and Technical Systems

National Research Council

NATIONAL ACADEMY PRESS
Washington, D.C. 1996

NATIONAL ACADEMY PRESS • 2101 Constitution Avenue, N.W. • Washington, D.C. 20418

NOTICE: The project that is the subject of this report was approved by the Governing Board of the National Research Council, whose members are drawn from the councils of the National Academy of Sciences, the National Academy of Engineering, and the Institute of Medicine. The members of the panel responsible for the report were chosen for their special competencies and with regard for appropriate balance.

This report has been reviewed by a group other than the authors according to procedures approved by a Report Review Committee consisting of members of the National Academy of Sciences, the National Academy of Engineering, and the Institute of Medicine.

The National Academy of Sciences is a private, nonprofit, self-perpetuating society of distinguished scholars engaged in scientific and engineering research, dedicated to the furtherance of science and technology and to their use for the general welfare. Upon the authority of the charter granted to it by the Congress in 1863, the Academy has a mandate that requires it to advise the federal government on scientific and technical matters. Dr. Bruce Alberts is president of the National Academy of Sciences.

The National Academy of Engineering was established in 1964, under the charter of the National Academy of Sciences, as a parallel organization of outstanding engineers. It is autonomous in its administration and in the selection of its members, sharing with the National Academy of Sciences the responsibility for advising the federal government. The National Academy of Engineering also sponsors engineering programs aimed at meeting national needs, encourages education and research, and recognizes the superior achievements of engineers. Dr. Harold Liebowitz is president of the National Academy of Engineering.

The Institute of Medicine was established in 1970 by the National Academy of Sciences to secure the services of eminent members of appropriate professions in the examination of policy matters pertaining to the health of the public. The Institute acts under the responsibility given to the National Academy of Sciences by its congressional charter to be an adviser to the federal government and, upon its own initiative, to identify issues of medical care, research, and education. Dr. Kenneth I. Shine is president of the Institute of Medicine.

The National Research Council was organized by the National Academy of Sciences in 1916 to associate the broad community of science and technology with the Academy's purposes of furthering knowledge and advising the federal government. Functioning in accordance with general policies determined by the Academy, the Council has become the principal operating agency of both the National Academy of Sciences and the National Academy of Engineering in providing services to the government, the public, and the scientific and engineering communities. The Council is administered jointly by both Academies and the Institute of Medicine. Dr. Bruce Alberts and Dr. Harold Liebowitz are chairman and vice-chairman, respectively, of the National Research Council.

The program described in this report is supported by cooperative agreement No. 14-35-0001-30475 between the Minerals Management Service of the U.S. Department of the Interior and the National Academy of Sciences and by interagency cooperative agreement No. DTMA91-94-G-00003 between the U.S. Maritime Administration of the U.S. Department of Transportation and the National Academy of Sciences.

Limited copies are available from:

Marine Board
Commission on Engineering and Technical Systems
National Research Council
2101 Constitution Avenue, N.W.
Washington, D.C. 20418

Additional copies are available for sale from:

National Academy Press
2101 Constitution Avenue, N.W.
Box 285
Washington, D.C. 20055
800-624-6242
(202) 334-3313 (in the Washington Metropolitan Area)

Library of Congress Catalog Card Number 95-72455
International Standard Book Number 0-309-05383-8

Copyright © 1996 by the National Academy of Sciences. All rights reserved.

Cover Photo: STAR Center, Dania, Florida

Printed in the United States of America

COMMITTEE ON SHIP-BRIDGE SIMULATION TRAINING

WILLIAM A. CREELMAN, *Chair*, Marine Consultant, Cape Cod, Massachusetts
PETER BARBER, The Southampton Institute, United Kingdom
ANITA D'AMICO BEADON, Northrop Grumman Corporation, Beth Page, New York
PETER H. CRESSY, University of Massachusetts, Dartmouth
DOWARD G. DOUWSMA, Grafton Group, Gainesville, Georgia
PHYLLIS J. KAYTEN, Federal Aviation Administration, Moffett Field, California
GAVAN LINTERN, University of Illinois, Savoy
DANIEL H. MacELREVEY, Consultant, Wildwood Crest, New Jersey
EDMOND L. MANDIN, Consultant, Kentfield, California
ROBERT J. MEURN, U.S. Merchant Marine Academy, Kings Point, New York
J. NICHOLAS NEWMAN, **NAE,** Massachusetts Institute of Technology, Cambridge
RICHARD A. SUTHERLAND, Consultant, Annandale, Virginia

Government Liaisons

LARRY DAGGETT, U.S. Army Corps of Engineers, Vicksburg, Mississippi
ALEXANDER C. LANDSBURG, U.S. Maritime Administration, Washington, D.C.
CHRISTINE S. MEERS, U.S. Coast Guard, Washington, D.C.

Staff

WAYNE YOUNG, Senior Staff Officer (until January 1995)
MARLENE R.B. BEAUDIN, Senior Staff Officer (from January 1995)
RICKY PAYNE, Administrative Assistant (until March 1994)
CARLA D. MOORE, Administrative Assistant (from January 1995)

MARINE BOARD

RICHARD J. SEYMOUR, *Chair*, Texas A&M University and Scripps Institution of Oceanography, La Jolla, California
BERNARD J. ABRAHAMSSON, University of Wisconsin, Superior
JERRY A. ASPLAND, (retired) ARCO Marine, Inc., Long Beach, California
ANNE D. AYLWARD, Volpe National Transportation Systems Center, Cambridge, Massachusetts
MARK Y. BERMAN, Amoco Corporation, Tulsa, Oklahoma
BROCK B. BERNSTEIN, EcoAnalysis, Ojai, California
JOHN W. BOYLSTON, Argent Marine Operations, Inc., Solomons, Maryland
SARAH CHASIS, Natural Resources Defense Council, Inc., New York, New York
CHRYSSOSTOMOS CHRYSSOSTOMIDIS, Massachusetts Institute of Technology, Cambridge
BILIANA CICIN-SAIN, University of Delaware, Newark
JAMES M. COLEMAN, **NAE**, Louisiana State University, Baton Rouge
BILLY L. EDGE, Texas A&M University, College Station
MARTHA GRABOWSKI, LeMoyne College and Rensselaer Polytechnic Institute, Cazenovia, New York
M. ELISABETH PATÉ-CORNELL, **NAE**, Stanford University, Stanford, California
DONALD W. PRITCHARD, **NAE**, State University of New York at Stony Brook, Severna Park, Maryland
STEPHANIE R. THORNTON, Coastal Resources Center, San Francisco, California
KARL K. TUREKIAN, **NAS**, Yale University, New Haven, Connecticut
ROD VULOVIC, Sea-Land Service, Inc., Elizabeth, New Jersey
E.G. "SKIP" WARD, Shell Offshore, Inc., Houston, Texas
ALAN G. YOUNG, Fugro-McClelland BV, Houston, Texas

Staff

CHARLES A. BOOKMAN, Director
DONALD W. PERKINS, Associate Director
DORIS C. HOLMES, Staff Associate

Acknowledgments

The committee is indebted to many individuals and organizations who generously made presentations and provided information and access to facilities during the course of its work.

We first thank our three sponsors' liaisons, Alexander C. Landsburg of the U.S. Maritime Administration, Christine Meers of the U.S. Coast Guard, and Dr. Larry Daggett of the U.S. Army Corps of Engineers, who provided helpful and expert counsel throughout our study.

The committee visited a number of facilities during its work. One such facility was the Computer Assisted Operations Research Facility (CAORF) at the U.S. Merchant Marine Academy, Kings Point, New York. Rear Admiral Paul Krinsky, Superintendent of the Academy, and Dr. F. Eugene Guest of MarineSafety International, operators of CAORF, and their staffs were particularly helpful. Another simulator facility visited was Maritime Institute of Technology and Graduate Studies at Linthicum, Maryland, where our host and guide was Captain Charles Pillsbury and whose expert staff was most helpful.

The Seaman's Church Institute simulator facility in New York City, under the direction of Captain Richard Beadon, was also toured by the committee. Subsequently, a subcommittee was generously afforded the opportunity to experience a simulator-based ship-bridge team training course at this facility.

The committee appreciates the hospitality of the American Marine Officer's (AMO) Raymond T. McKay Simulation Training and Research (STAR) Center at Dania, Florida, under the direction of Captain Harry Crooks of AMO and Tom Mara of SIMSHIP Corporation. While touring this facility, the committee observed simulation training and held a number of meetings. Captain

Frank Seitz and Brian Long of SIMSHIP, as well as members of their staffs, were extremely helpful.

Other facilities that welcomed and briefed the committee included the following: (1) the Naval Air Warfare Center, Training Systems Division in Orlando, Florida, where the committee was hosted by Commander McMeekan; (2) the Naval Amphibious School in Little Creek, Virginia, where committee members rode manned models and observed radio-controlled models in a maneuvering basin; (3) the Delta Airlines Training Facility in Atlanta, Georgia, where each committee member experienced the 767 aircraft simulators; (4) the U.S. Naval Academy in Annapolis, Maryland, where the committee attended Blake Bush's interactive rules-of-the-road class and used the Academy's ship-bridge simulator; and (5) the Massachusetts Maritime Academy, where the committee chair toured the simulator facilities and observed simulator training of cadets.

The committee greatly appreciates insightful presentations made in Atlanta by Captain Ed Moir, systems manager, Delta Flight Training; Captain Tony Papandrea, manager, Delta Flight Training Operations; Paul Ray of the Federal Aviation Administration National Flight Simulation Program; Thomas Longridge of the Federal Aviation Administration; and Phil Buscovitch, manager, Advanced Qualification Program, Delta Airlines.

Many individual ships' pilots and pilot associations gave generously of their time and expertise. Among them were Captain Jack Sparks and Paul Kirschner of the American Pilots' Association; Vincent Black and Captain H.W. Mahlmann of the United Sandy Hook Pilots of New York and New Jersey; Captain Richard Owen of the Association of Maryland Pilots; Captain Edward Cray of the Port Everglade Pilots; and Captain S. Orlando Allard, chief, maritime training, the Panama Canal Commission.

Technical presentations were also made to the committee by Thomas Hammell of Paradigm Associates; Wei-Yang Hwang of MarineSafety International; Rear Admiral Floyd H. Miller, superintendent of the New York Maritime Academy; George Sandberg of the U.S. Merchant Marine Academy; Dean Albert Higgins of the Maine Maritime Academy; and Jim Brown of the Harry Lundeberg School of Seamanship. Presentations on desktop and part-task simulations were demonstrated by Franklin Gaines of the Mariners' License Preparation School and Greg Szczurek of Examco and Houston Marine. Each of these people contributed to the knowledge and understanding of the committee.

Vessel operators and their consultants shared their valuable insights about simulator training with the committee. Among these were James Sanborn of Maritrans; James Sweeney of Morania Oil Tanker Corporation and their consultant, Captain Herb Groh; John Boylston of Argent Marine; Captain Saunders Jones of American President Lines; and Morris Croce of Chevron Shipping.

The committee also benefited greatly from the efforts of committee member Captain Edmond Mandin, who arranged rides for committee members from San Pedro, California, to Oakland aboard American President Line containerships

President Lincoln and *President Roosevelt*. The officers and crews of those vessels were extremely helpful to the committee. Committee member Robert Meurn also arranged lodging, tours of simulator facilities, and use of meeting facilities at U.S. Merchant Marine Academy and CAORF, which are most appreciated.

The committee is grateful to Wayne Young, staff director from the inception of the study in 1992 through year-end 1994, for his efforts in arranging and coordinating the presentations at our many meetings and to Marlene Beaudin, staff director following Wayne's departure, for the intensive editing and organizational work she so ably performed. Our appreciation goes to the staff work of Ricky Payne who supported our work until his illness in 1994 and to Carla Moore who succeeded him.

The committee has been the beneficiary of extraordinary cooperation and helpfulness from all areas of relevant expertise—vessel operators, maritime and aviation training experts, learning system experts, simulator operators, simulation software designers, maritime and aviation licensing officials, and ships' officers and pilots. Their generous assistance made this study possible.

Preface

BACKGROUND

The professional performance of merchant mariners, marine pilots, and towing vessel operators has been brought to public attention by major marine accidents and the resulting loss of life, oil spills, and damage to marine ecosystems. Intense public and congressional interest in the safety of marine transportation and the qualification of deck officers and pilots was stimulated by the *Exxon Valdez* grounding and oil spill in Prince William Sound, Alaska, in 1989. National interest has been sustained by newsworthy marine accidents along the nation's seacoasts involving all major categories of commercial vessels—cargo ships, tankers, coastwise and inland towing vessels, and passenger vessels. Faulty human performance has figured prominently in most of these accidents.

The U.S. Congress, in the Oil Pollution Act of 1990 (OPA 90 P.L. 101-380), recognized the potential value of simulation in training deck officers and pilots of vessels transporting oil and hazardous substances in U.S. waters. The act directed the U.S. Coast Guard (USCG) to determine the feasibility and practicality of mandating simulator-based training.

Among the recommendations of a subsequent USCG internal study, *Licensing 2000 and Beyond* (Anderson et al., 1993), were recommendations that the agency do the following:

- assess computer-based training and testing systems for possible application to mariner training and licensing, and
- conduct research and development for microcomputer desktop training programs.

PREFACE

Marine simulation in various forms is not new to the marine community. Ship maneuvering simulations, manned models, and radar simulations have been used for specialized training in the marine community since the 1960s. The U.S. Maritime Administration (MarAd) and the USCG conducted extensive operational research into human performance during the 1970s and early 1980s using a computer-based ship-bridge simulator constructed for that purpose: the Computer Aided Operations Research Facility (CAORF) located on the grounds of the U.S. Merchant Marine Academy in Kings Point, New York. Although this research diminished significantly after the privatization of the CAORF facility in the mid-1980s, the published research, albeit somewhat dated, still serves as a major resource for guiding the use of marine simulation in professional development.

Over the past decade, labor unions and private companies have made substantial investments worldwide in the establishment of computer-based marine simulator facilities. These facilities have been used extensively by some shipping and towing companies and marine pilot associations for specialized training and continuing professional development. Radar simulation is now used extensively for prerequisite radar observer training as part of marine licensing and certification requirements worldwide. The port community, including port authorities and marine pilots, is also familiar with shiphandling simulation through its application to channel design.

Given this background, the USCG requested that the National Research Council of the National Academy of Sciences conduct a comprehensive assessment of the role of ship-bridge simulation in professional development and marine licensing. In requesting the study, the agency indicated that the report should focus on the use of ship-bridge simulation in training U.S. deck officers and pilots of all categories of ships operating in U.S. coastal and harbor waters, as well as operators of coastwise towing vessels. The agency indicated that other forms of marine simulators—such as manned models, limited-task simulators, and desktop computer-based simulators—should be included for comparative purposes. The use of simulation for inland towing vessels is beyond the scope of this assessment.

FORMATION OF THE
NATIONAL RESEARCH COUNCIL COMMITTEE

The National Research Council convened the Committee on Ship-Bridge Simulation Training under the auspices of the Marine Board of the Commission on Engineering and Technical Systems. Committee members were selected for their expertise and to ensure a wide range of experience and viewpoints. Consistent with the policy of the National Research Council, the principle guiding constitution of the committee and its work was not to exclude a potential member with expertise vital to the study because of potential biases, but rather to seek a balance among the members of the committee as a whole.

Committee members were selected for their experience in marine pilotage, hydrodynamics, marine and adult education theory and practice, marine and aviation simulation, ship and towing vessel operations and management, ship-handling, marine licensing, and human performance. Academic, industrial, government, and engineering perspectives were reflected in the committee's composition. Brief biographies of committee members are provided in Appendix A.

The committee was assisted by the USCG, MarAd, and U.S. Army Corps of Engineers, each of which designated liaison representatives.

SCOPE OF STUDY

The committee was asked to conduct a multidisciplinary assessment of the role of ship-bridge simulation in the professional development and licensing of mariners responsible for vessel navigation and piloting. Included in the scope of study were:

- the state of practice in applying ship-bridge simulation technology for pilot, individual, and team training;
- the potential, role, and cost-effectiveness[1] of ship-bridge simulation for initial maritime education, licensing, skills development, skills maintenance, indoctrination to emerging navigation and ship technologies, and determining professional competence;
- the scientific, technical, and professional bases for substituting simulation for practical experience in professional development and in marine licensing;
- requirements for the development and validation of ship-bridge simulators and simulations; and
- improvements and research needed in technology and applications to advance the state of practice of ship-bridge simulation for professional development and marine licensing.

The committee was asked to focus on computer-based, full-mission and limited-task simulators, full-bridge mockups, and the operation of oceangoing and coastwise vessels in coastal waters, harbors, and waterways. Manned models, desktop simulators, virtual reality systems, and other forms of marine simulators are included as alternatives for comparative analysis.

COMMITTEE ACTIVITIES

As a part of its background research, the committee commissioned a literature search that initially identified approximately 1,400 possibly relevant

---

[1] The committee was not able to make a detailed analysis of the cost-effectiveness of ship-bridge simulation. Chapter 2 does, however, contain some information on the range of cost of simulator-based training.

PREFACE   xi

documents. From that list, 103 of the most relevant documents were selected and abstracted for the committee (Douwsma, 1993). Of that number, the committee found that nearly half were technical papers and reports of research and experimentation from CAORF (primarily from the 1970s and early 1980s), and half were conference and journal articles.

During the course of the study, the committee also visited a number of marine and one air carrier simulation facilities, including:

- CAORF, Kings Point, New York;
- Maritime Institute of Technology and Graduate Studies, Maryland;
- Seaman's Church Institute, New York;
- STAR Center, Florida;
- Naval Air Warfare Center, Florida;
- Naval Amphibious School, Virginia;
- U.S. Naval Academy, Maryland;
- Massachusetts Maritime Academy, Massachusetts; and
- Delta Airlines Training Facility, Georgia.

In addition, the committee received presentations from and participated in discussions with over 30 experts, one of which was a detailed orientation on the experiences of Panama Canal pilots by Captain S. Orlando Allard (see Allard, 1993). Several members of the committee also attended a three-day simulator-based ship-bridge team training course at Seaman's Church Institute.

REPORT ORGANIZATION

This report was prepared for state and federal government decision makers, marine safety and licensing authorities, mariners, the shipping and coastwise towing industries, marine education and training facilities, and public interest organizations.

Moreover, the USCG, on behalf of the United States, has joined a number of other delegations to the International Maritime Organization in urging international requirements and standards for the use of simulation in qualifying mariners. A major revision of the international standards that establish a baseline for knowledge and skill requirements leading to the issuance of marine licenses for seagoing ships is in progress. The report could serve as a resource in addressing and implementing changes in these areas. Chapter 1 reviews the duties and responsibilities of deck officers and marine pilots and introduces issues in the professional development of mariners and the international and national context for mariner training, licensing, and professional development. Chapter 2 describes the classification and types of simulators and summarizes the current state of practice in the use of simulators for training and licensing. This chapter also reviews the use of simulators by the commercial air carrier industry. Chapter 3 discusses the instructional design process and identifies the key elements that comprise an effective training program. Chapter 4 describes the training environment produced by computer-based simulators and by manned models and relates that environment to training objectives discussed in Chapter 3. Chapter 5 discusses the use of simulation for evaluation of training performance and for assessment of performance in marine licensing. Chapter 6 examines the practice and appropriateness of using marine simulator-based training as a substitute for seagoing service to meet international standards for marine certification. Chapter 7 discusses the validity, validation, and assessment of simulators and simulation. Chapter 8 presents the committee's conclusions and recommendations. Appendix A contains brief biographical sketches of committee members. Appendices B–G provide essential background and contextual and technical information related to the analyses in the main body of the report.

REFERENCES

Allard, S.O. 1993. Pilot Training at the Panama Canal. Unpublished paper for the Committee on Ship-Bridge Simulation Training. Marine Board, National Research Council, Washington, D.C.

Anderson, D.B., T.L. Rice, R.G. Ross, J.D. Pendergraft, C.D. Kakuska, D.F. Meyers, S.J. Szczepaniak, and P.A. Stutman. 1993. Licensing 2000 and Beyond. Washington, D.C.: Office of Marine Safety, Security, and Environmental Protection, U.S. Coast Guard.

Douwsma, D.G. 1993. Background Paper: Shiphandling Simulation Training. Unpublished literature review prepared for the Committee on Ship-Bridge Simulation Training, Marine Board, National Research Council, Washington, D.C.

Contents

	EXECUTIVE SUMMARY	1
1	MARINERS—THEIR WORK AND PROFESSIONAL DEVELOPMENT	13
	The Mariner Population, 14	
	U.S. and International Operating Environments, 20	
	Mariner Training, Licensing, and Professional Development, 22	
	References, 35	
2	USE OF SIMULATION IN TRAINING AND LICENSING: CURRENT STATE OF PRACTICE	37
	Rationale for Using Simulators, 37	
	Types of Simulators, 38	
	Use of Marine Simulators for Training, 51	
	Simulation in Marine Licensing, 56	
	Use of Simulation in Nonfederal Marine Licensing, 59	
	Cost of Simulator-Based Training, 60	
	Simulation in the Commercial Air Carrier Industry, 61	
	References, 65	
3	EFFECTIVE TRAINING WITH SIMULATION: THE INSTRUCTIONAL DESIGN PROCESS	67
	Developing an Effective Training Program, 67	
	Applying Instructional Design, 71	

Determining Training Methods, 74
Transfer and Retention of Training, 81
Measuring Training Program Effectiveness, 83
Simulator-Based Training Instructors, 85
Findings, 93
References, 95

4 MATCHING THE TRAINING ENVIRONMENT TO OBJECTIVES 97
Establishing Levels of Simulation, 98
Relative Importance of Simulator Components, 98
Simulator Components and Training Objectives, 100
Findings, 118
References, 118

5 PERFORMANCE EVALUATION AND LICENSING ASSESSMENT 120
Understanding Performance Evaluation and Assessment, 120
Forms of Evaluation and Assessment, 122
Training and Evaluation with Simulators, 126
Licensing Performance Assessment with Simulators, 130
Issues in Simulation Evaluation or Assessment, 131
Findings, 140
References, 142

6 SIMULATOR-BASED TRAINING AND SEA-TIME EQUIVALENCY 143
International Sea-Time Requirements, 143
Definition of Sea-Time Equivalency, 144
Sea-Time Equivalency and Mariner Competency, 145
Basis for Sea-Time Equivalency, 146
Issues Affecting Sea-Time Equivalency Decisions, 147
A Systematic Approach to Determining Sea-Time Equivalency, 150
Possible Equivalency Applications, 152
Findings, 155
References, 156

7 SIMULATION AND SIMULATOR VALIDITY AND VALIDATION 158
The Fidelity-Accuracy Relationship, 159
Physical and Mathematical Simulation Models, 160
Current Practice in Validation, 163
Facility-Generated Models and Modifications, 164
An Approach to Simulator and Simulation Validation, 165
Issues and Future Developments, 169

    Findings, 170
    References, 171

8   CONCLUSIONS AND RECOMMENDATIONS   173
    Use of Simulators for Training, 174
    Use of Simulators in the USCG Licensing Program, 176
    Validation of Simulations and Simulators, 180
    Research Needed to Improve Mariner Training, Licensing, and
       Professional Development, 181
    Funding Simulator-Based Training and Licensing, 184

APPENDICES

    A Biographical Sketches of Committee Members   189

    B International Marine Certification Roles, Responsibilities, and
       Standards   195

    C Professional Licensing Infrastructure for U.S. Merchant Mariners   204

    D Hydrodynamics, Physical Models, and Mathematical Modeling   211

    E Outlines of Sample Simulator-Based Training Courses   241

    F Uses of Simulators: Illustrative Case Studies   246

    G Microcomputer Desktop Simulation   272

LIST OF BOXES, FIGURES, AND TABLES

BOXES

1-1 Selected Definitions, 15
1-2 Mariner Professional Development: Training and Licensing, 23
1-3 The National Vocational Qualification (NVQ) System of the United Kingdom, 34
2-1 Marine Operations Bridge Simulators Classifications Proposed to International Maritime Organization (IMO), 39
2-2 Location of U.S. Facilities with Category I and Category II Simulators, 42
3-1 Elements of Instructional Design Process, 70
3-2 Training Insights from Mariner Instructors, 72
3-3 Instructional Tasks, 87
3-4 Samples of Instructor Training Programs, Maritime Academy Simulator Committee (MASC): Draft "Train-the-Instructor" Course, 89
3-5 Samples of Instructor Training Programs, The Southampton Institute, Warsash Maritime Centre, United Kingdom, Full-Mission Ship-Bridge Simulator, 90
3-6 Samples of Instructor Training Programs: MarineSafety International Rotterdam, 92
4-1 Computer-Generated Image (CGI) Projection Systems, 106
5-1 Use of Simulators for Performance Evaluation: The Panama Canal Commission, 130
5-2 Comments on Testing Pilots Using Simulators, 131
5-3 Typical Summary of a Simulator-Based Check-Ride, 132
7-1 Anchoring Evolutions: An Example of Needed Research, 161
F-1 Third Mate Observations on Value of Ship-Bridge Simulation Cadet Watchkeeping Course, U.S. Merchant Marine Academy, 258
F-2 Observations of Panama Canal Pilots on the Value of Ship-Bridge Simulation Training, 264
G-1 Testing Objectives for Mariner License Testing Devices, 278
G-2 Development Criteria for U.S. Coast Guard License Testing Devices, 279

FIGURES

2-1 Types of Marine Simulators, 41
2-2 View of the Bridge of a Full-Mission Simulator, 44
2-3 View of the Bridge of a Full-Mission Simulator, 45
2-4 Elements of a Sample PC-Based Simulator Program, 46
2-5 Elements of a Sample PC-Based Simulator Program, 47
2-6 An Example of a Manned-Model Simulator, 49
2-7 An Example of a Manned-Model Simulator, 50

3-1	The Training Process, 69
4-1	Levels of Sophistication for Simulator Physical Components, 99
4-2	Control and Monitoring Station, 101
4-3	Estimate of Relative Importance of Ship-Bridge Equipment for Simulator Training, 102
D-1	Paths of Stable and Unstable Ships after a Yaw Disturbance of 1 Degree, 218
D-2	Steady Turning Rate Versus Rudder Angle, 220
D-3	Spiral Test, 221
D-4	Zig-zag Maneuver Response: 5–5 Degree, 222
F-1	Cadet Watch Team Grading Sheet, 252
F-2	Cadet Watch Team Evaluation Sheet, 253
F-3	Scores Achieved by 233 Cadet Watchkeeping Teams Undergoing a Simulation-Based Watchkeeping Course During the Period 1985–1994, 255
F-4	Plot of Standard Deviation for Simulation-Based Cadet Watchstanding Course, 256
F-5	Average and Weighted Average Scores Per Drill of Simulation-Based Watchkeeping Training at the U.S. Merchant Marine Academy, 257

TABLES

1-1	Historical Inventory of World and U.S.-Flag Ocean Ships Over 1,000 Gross Tons, 21
1-2	U.S. Coast Guard Ocean-Only License Statistics for Deck Department, Any Gross Tons, Fiscal Years 1986–1993, 30
1-3	U.S. Coast Guard Limited and Unlimited License Activity and Number of Facilities with Category I and Category II Simulators: Summarized by Region, 31
1-4	U.S. Coast Guard Total Limited and Unlimited Licenses, by Category, 1994, 32
D-1	Principal Particulars, 219

Executive Summary

Concern about recent marine accidents has focused attention on the use of simulators for the training, performance evaluation, and licensing assessment of mariners. Over the past decade, labor unions, private companies, and the U.S. government have made substantial investments worldwide in the development and use of simulator-based training facilities. This report by the Committee on Ship-Bridge Simulation Training is an assessment of the role of ship-bridge simulation in the professional development of U.S. deck officers (masters, chief mates, and mates), cadets, and marine pilots.

MARINERS AND PROFESSIONAL DEVELOPMENT

Mariners' duties and responsibilities are dictated by their work environment. Deck officers must be generalists, operating in the sometimes highly stressful and demanding environment of automated ships, short turnaround times in port, smaller crew sizes, and the self-contained independence of long sea voyages. Deck officers must be knowledgeable in skills ranging from watchkeeping, navigation, cargo handling, and radar to medical care, life saving, maritime law, and ship's business.

Marine pilots are also highly skilled generalists who function independently in an environment that requires them to understand the operation of ship-bridge equipment and maneuvering capabilities of a wide range of vessels and to be able to safely maneuver through shallow and restricted waters. Pilots must also be knowledgeable in local customs and working practices of ports and terminal operations.

Historically, the professional development of mariners has been based on a strong tradition of on-the-job learning. Graduate mariners supplement training garnered in the structured academic environment with colleague-assisted training and occasional structured training courses at union or other instructional facilities. Mariners who become marine pilots advance through apprenticeship programs.

OVERVIEW OF THE USE OF SIMULATORS

There is a wide range of marine simulators in use worldwide. The capabilities of computer-based simulators range from radar only to full-scale ship-bridge simulators capable of simulating a 360-degree view. Marine simulators can simulate a range of vessels in scenarios of real or generic operating conditions (e.g., ports and harbors). Computer-based simulators can be used to train mariners in a number of skills, from rules of the road and emergency procedures to bridge team and bridge resource management.

Physical scale-model, or manned-model, simulators are scale models of specific vessels that effectively simulate ship motion and handling in fast time. These models are especially effective for teaching shiphandling and maneuvering skills.

As a training tool, simulators have a number of significant advantages:

- Simulators can be used to train regardless of weather.
- Instructors can terminate training scenarios at any time.
- Training scenarios can be repeated.
- Training scenarios can be recorded and played back.
- Training takes place in a "safe" environment.

In applying simulation to training requirements, it is important to consider differences among simulators, that is, the different levels of simulator component capabilities. A high degree of realism is not always required for effective learning transfer. Often it is not necessary to use the most sophisticated simulator to meet all training objectives. However, the levels of realism and accuracy required should match the training objectives.

Ship-bridge simulators are different from commercial air carrier simulators. Commercial air carrier simulators are airframe-specific and are subject to industrywide standards. The simulators are validated and periodically revalidated (every four months) by the National Simulator Evaluation Program.

There are no industrywide validation programs for marine simulators. Ship-bridge simulators usually simulate a variety of real or generic ship types—from coastwise tugs to very large crude carriers.

Simulators are also used for mariner performance evaluation. These evaluations are usually informal and take the form of debriefings during the course of training. Occasionally, however, simulators are used for more structured evaluations. A number of the cadet training programs at maritime academies use them extensively for performance evaluation.

Currently, simulators are used in two U.S. Coast Guard (USCG) licensing programs to:

- demonstrate knowledge of fundamentals of radar in qualifying for radar observer certification, and
- receive the unlimited master's license by meeting standard prerequisites and successfully completing the training course offered at the STAR Center facility in Florida.

SUMMARY OF CONCLUSIONS AND RECOMMENDATIONS

The Committee on Ship-Bridge Simulation Training found that simulation can become an effective training tool to improve mariner professional competence. It also found that simulation offers the USCG licensing program a mechanism for determining whether mariners are competent on a much more comprehensive basis than through its current written multiple-choice examinations. However, for the USCG to use simulation effectively for training and licensing, it is important that a stronger research base be developed and that the agency address issues of standardization and validation discussed in this report. The following is a summary of the committee's conclusions and recommendations. The complete text is included in Chapter 8.

Use of Simulators for Training

Setting Standards for Simulator-Based Training Courses

Mariner training is strongly task-oriented. Many of the current approaches to training and professional development in the marine industry have been based on a tradition of "modeling the expert" and on-the-job training. Training programs using simulation often insert simulation into existing courses rather than customizing the course to ensure that the simulation contributes effectively to the course training objectives. One result has been a lack of standardization in simulator-based courses.

The Committee on Ship-Bridge Simulation Training believes that the greatest benefits of simulation will be realized with a more structured approach to the use of simulation for training and with the standardization of some of the key courses of instruction. Systematic application of the instructional design process offers a strong model for the structuring of new courses and the continuous improvement of existing courses.

The committee also found that the instructor can be more important than the simulation in meeting training objectives. It is the instructor's responsibility to ensure that all training objectives are met. Instructors who train on simulators should themselves be trained and certified to ensure that the simulator-based training courses meet the training objectives.

Recommendation 1: Marine simulation should be used in conjunction with other training methodologies during routine training, including cadet training at the maritime academies, for the development and qualification of professional mariner knowledge and skills.

Recommendation 2: The U.S. Coast Guard should oversee and guide the establishment of nationally applied standards for all simulator-based training courses within its jurisdiction. Standards development should include consultation with, and perhaps use of, outside expertise available in existing advisory committees, technical groups, forums, or special oversight boards. If the USCG relies on outside bodies, the process should be open and include interdisciplinary consultation with professional marine, trade, labor, and management organizations; federal advisory committees; professional marine pilot organizations; and marine educators, including state and federal maritime academies. Whatever process the USCG chooses to use should be acceptable from a regulatory standpoint.

Recommendation 3: U.S. licensing authorities should require that instructors of simulator-based training courses used for formal licensing assessment, licensing renewal, and training for required certifications (i.e., liquid natural gas carrier watchstander, offshore oil port mooring masters) be professionally competent with respect to relevant nautical expertise, the licensing process, and training methods. The professional qualifications of the lead instructor should be at least the same as the highest qualification for which trainees are being trained or examined. Criteria and standards for instructor qualification should be developed and procedures set in place for certifying and periodically recertifying instructors who conduct training.

Use of Simulators to Promote Continuing Professional Development

The Committee on Ship-Bridge Simulation Training found that computer-based ship-bridge simulators and manned models can be effective in the development and renewal of deck officers and marine pilot skills in a number of significant areas including bridge team management and bridge resource management, shiphandling, docking and undocking evolutions, bridge watchkeeping, rules of the road, and emergency procedures. Although current computer-based simulators are limited in their ability to simulate ships' maneuvering trajectories in shallow and restricted waterways and ship-to-ship interactions—capabilities important to pilot shiphandling training—computer-based simulator training in areas such as bridge team management and bridge resource management can be of value to pilots.

The increase in the number of foreign-flag vessels in U.S. waters has given simulator-based training a global dimension. The International Maritime Organization (IMO) is currently making extensive revisions to its Standards for Training, Certification, and Watchkeeping guidelines, including guidance for use of simulators. To effectively reduce marine accidents in U.S. waters it will be important that international training standards be developed.

Recommendation 4: Marine pilotage authorities and companies retaining pilot services should encourage marine pilots, docking masters, and mooring masters who have not participated in an accredited ship-bridge simulator or manned-model course to do so as an element of continuing professional development. Marine pilot organizations, including the American Pilots' Association and state commissions, boards, and associations, should, in cooperation with companies retaining pilot services, establish programs to implement this recommendation.

Recommendation 5: Use of simulators for professional development should be implemented on an international scale to enhance the professional development of all mariners who operate vessels entering U.S. waters and to reduce the potential for accidents. The USCG should advocate this strategy in its representation of marine safety interests to the IMO and other appropriate international bodies.

Special-Task Simulators

Special-task or microcomputer simulators could be used effectively in mariner training. A limitation affecting the widespread use of microcomputers is the limited availability of desktop simulations and interactive courseware.

Recommendation 6: The U.S. Department of Transportation should selectively sponsor development of interactive courseware with embedded simulations that would facilitate the understanding of information and concepts that are difficult or costly to convey by conventional means.

Use of Simulators in the USCG Licensing Program

Need for a Plan to Use Simulators Effectively

Most USCG license examinations test knowledge through structured, objective, multiple-choice examinations. These examinations are, in many cases, dated. The use of simulation offers an effective mechanism for accessing not only the candidate's knowledge but also his or her ability to apply that knowledge, to prioritize tasks, and to perform several tasks simultaneously, all functions routinely required aboard ship. For the USCG to effectively use simulation for testing of skills, abilities, and multiple tasks, the agency needs an understanding of the nature of those tasks and a detailed plan for effective, structured assessment. It is therefore important that the USCG develop a framework for integrating simulation into its marine licensing program before it undertakes more extensive use of simulation in marine licensing.

The quality of the license assessors is another issue the USCG should carefully consider. Assessors should be trained and certified. In cases where a simulator-based training course is integrated into the licensing process, the instructor and assessor must be two different people.

The USCG will need to factor into its plans sufficient time to not only allow new simulator facilities to come on-line but also to allow sufficient time for the appropriate bodies to develop standards and establish a process for validation of simulators and simulations (Recommendations 14 and 15)

Recommendation 7: The U.S. Coast Guard should develop a detailed plan to restructure its marine licensing program to incorporate simulation into the program and to use simulation as a basis of other structured assessments. In development of the plan, the USCG should consult with all parties of interest.

Recommendation 8: Licensing authorities should require that license assessors of simulator-based licensing examinations be professionally competent with respect to relevant nautical expertise, the licensing process, and assessment methods. Assessors should hold a marine license at least equal to the highest qualification for which the candidate is being tested or should be a recognized expert in a specialized skill being trained. Specific criteria and standards for assessor qualification should be developed, and procedures should be set in place for certifying and periodically recertifying assessors who conduct licensing assessments with simulators.

Substitution of Simulator Training for Required Sea Service

The USCG currently grants remission of required sea time in some ratio of sea time to training time for successful completion of specified simulator-based training courses. The current program has evolved ad hoc rather than through systematic technical analysis. Cadets, for example, are granted remission of sea time in a 6 to 1 ratio for the successful completion of a watchstanding simulator-based training course. There was no research basis for the use of that ratio. The USCG has stated that it wishes to expand its remission policy as a mechanism for encouraging the use of simulation-based training.

The Committee on Ship-Bridge Simulation Training concluded that, in some cases, remission of sea time has the potential to compromise mariner competency and safety and should therefore be approached cautiously. The USCG presently grants remission for license upgrades for approved full-mission simulator courses. Although there are areas where remission can be very effective (e.g., cadet training and license renewals), there are areas where the committee concluded that it is not currently appropriate (e.g., upgrade to second mate, chief mate, or master, or for any type of marine pilot's license).

EXECUTIVE SUMMARY   7

To implement a safe, effective program, the USCG should include the planning of sea-time remission in its framework for restructuring its licensing program. The standards for simulator-based training courses should be considered in the development of a plan for allowing substitution of simulator-based training for required sea time in the limited cases where the committee finds such remission to be suitable.

Recommendation 9: The U.S. Coast Guard should grant remission of sea time for the third mate's license for graduates of an accredited, professional development program that includes bridge watchkeeping simulation. The ratio of simulator time to sea time should be determined on a course-by-course basis and should depend on the quality of the learning experience as it applies to prospective third mates, including the degree to which the learning transfers to actual operations. Research to establish a more formalized basis for these determinations should be implemented without delay (Recommendation 16).

Recommendation 10: The U.S. Coast Guard should establish standards for the use of marine simulation as an alternative to sea service for recency requirements for license renewal of deck officers and vessel operators. Remission of sea time should be granted for renewal purposes to individuals who have successfully completed an accredited and USCG-approved simulator-based training course designed for this purpose. The course should be of sufficient length and depth and include rules-of-the-road training, bridge team and bridge resource management, and passage planning. The ratio of simulator time to sea time should be determined on a course-by-course basis and should depend on the quality of the overall learning experience insofar as this learning transfers effectively to actual operations.

Use of Simulator-Based Training During License Renewal

Active mariners can benefit from structured training at each license level, and inactive mariners can refresh vital skills in a relatively short time through simulator-based training. In both cases, simulation can play an important role by ensuring that applicants demonstrate some level of baseline competence at the time of license renewal.

Current operating practices in many segments of the U.S. merchant marine do not routinely provide adequate opportunities for chief mates to acquire essential shiphandling and bridge team and bridge resource management experience and expertise, much less proficiency. Development of these skills is very important. Simulation provides a safe, structured environment for learning these skills prior to receipt of a master's license and service as master.

Recommendation 11: Deck officers and licensed operators of oceangoing and coastwise vessels who can demonstrate recent shipboard or related experience, but who have not completed an accredited simulator-based training course, should be encouraged to complete an accredited simulator-based bridge resource or bridge team management course before their license renewal. Those seeking to renew licenses who cannot demonstrate recent shipboard experience should be required to complete such a course before returning to sea service under that license.

Recommendation 12: Chief mates should be required to complete an accredited hands-on shiphandling course prior to their first assignments as masters. The license should be endorsed to certify that training has been successfully completed. Either a manned-model or a computer-based, accredited shiphandling simulation course should be established as the norm for this training.

Recommendation 13: Currently serving masters who have not completed an accredited shiphandling simulation course should be required to do so prior to their next license renewal. In addition, masters should be encouraged to attend an accredited shiphandling simulator-based training course periodically thereafter.

Validation of Simulations and Simulators

The accuracy and fidelity of ship-bridge simulators can vary significantly from facility to facility. These differences derive from the differences among original mathematical models used to develop the simulations and from facility operator modifications to models after installation of the simulations. A number of facilities use in-house staff to develop their own models.

There are no industrywide simulator or simulation standards. Simulations are initially validated by the manufacturer, then by the facility operator through subjective tests. Often, facility operators periodically modify simulation models after the initial validation. This process of continually modifying simulation models can result in inconsistent training programs, as successive classes may be conducted with different simulations. These problems are of particular concern when a simulation is used for licensing or training related to remission of sea time.

To address these concerns, simulators and simulations should be validated, all modifications should be documented, and the simulation revalidated. The extent to which accuracy of a simulation needs to be validated will depend on the proposed use of the simulation.

Recommendation 14: The U.S. Coast Guard should enlist the assistance of standards-setting and other interested organizations and sponsor and support a structured process for validating and revalidating simulators, simulations, and assessment processes. In developing these standards, all parties of interest should be consulted.

Recommendation 15: Staff at simulator facilities should have objective knowledge of the capabilities and limitations of the hydrodynamic models on which their simulations are based. Modifications of the coefficients to address real or perceived deficiencies should only be performed based on competent oversight by a multidisciplinary team. Procedures should be developed to ensure that such changes are documented and that notification is given to the original vendors of the data and to the cognizant authorities.

Research Needed to Improve Mariner Training, Licensing, and Professional Development

Development of a Quantifiable Basis for Assessing Simulator Effectiveness

In most cases, the application of simulators for training and licensing has been supported by anecdotal information. The exact nature of the equivalency of simulation to real life, however, has not been systematically investigated for several reasons:

- the existing job-task analyses are not adequate for this purpose,
- there has been no systematic application of job-task analyses in either marine training or licensing for this purpose, and
- no systematic program currently exists to collect and analyze performance data for past participants in simulations.

The work of the mariner is task-oriented. The early research on task analysis was largely unused and is dated. To be able to effectively apply simulator technology, it is important to systematically measure and analyze simulator effectiveness for training and to develop a mechanism to use simulators to improve the effectiveness of the transfer of skills and knowledge.

Recommendation 16: The U.S. Coast Guard and U.S. Maritime Administration, in consultation with maritime educators, the marine industry, and the piloting profession, should sponsor a cooperative research program to establish a quantifiable basis for measuring the effectiveness of simulator-based training.

Simulation of the Physical Environment

Ship control and navigation are visually supported tasks, especially in confined areas. Learning visual skills is an important process in the development of proficiency in control and navigation. In many simulators, the visual simulations are provided with systems that have limited capabilities to represent the relative brightness of lights. The result can be distortion of distance perceptions as an observer moves around the simulated bridge.

The impact on training effectiveness of ship operational characteristics—such as vibration, sound, and physical movement of the bridge in roll, heave, and pitch—has not been verified and should be researched before applying these systems to simulators.

Recommendation 17: The U.S. Department of Transportation (DOT) and the maritime industry should assess the impact on training effectiveness of apparent limitations in simulator visual systems. If these limitations have a negative impact on training effectiveness, DOT should encourage development of visual systems that overcome or minimize the negative aspects of current systems.

Recommendation 18: The U.S. Department of Transportation should undertake structured assessments of the need for simulation of vibration, sound, and physical movement. These assessments should include consideration of the possibly differential value of these various sources of information in different types of training scenarios.

Manned Models

Manned models are an effective training device for illustrating and emphasizing the principles of shiphandling. They are particularly effective in providing hands-on ship maneuvering in confined waters, including berthing, unberthing, and channel work. Manned models can simulate more realistic representations of bank effects, shallow water, and ship-to-ship interactions than electronic, computer-driven ship-bridge simulators.

Recent closure of the U.S. Navy's manned-model facility at Little Creek, Virginia, which was one of only four such facilities in the world, represents the loss to the United States of a unique training resource. Although the contribution of this facility to operational safety of commercial vessels was a side benefit, it nevertheless filled a gap in U.S.-based training resources for the development of merchant mariners.

Recommendation 19: Because there are no manned-model training facilities in the United States, and because of the usefulness of these models in familiarizing pilots and others with important aspects of shiphandling, DOT should study the feasibility of establishing or re-establishing a manned-model shiphandling training facility in the United States, to be operated on a user-fee basis.

Vessel Maneuvering Behavior

The ability of a simulator to closely replicate a ship's maneuvering trajectory is a strong measure of the usefulness and value of the simulator for training and licensing. At present, computer-based simulation of a ship's maneuvering trajectory is well developed in normal deep-water, open-ocean cases. In cases involving shallow or restricted waterways, ship-to-ship interactions, and extreme maneuvers, fidelity may be significantly reduced.

Recommendation 20: The American Towing Tank Conference (ATTC) and International Towing Tank Conference (ITTC) should be advised of the needs to extend the database of ship maneuvering coefficients. ATTC and ITTC should be encouraged to investigate possible development of procedures that would allow the exploitation of existing proprietary data without source disclosure. Where data are not available from these sources, funds should be allocated to perform new tests, especially in very shallow water and in close proximity to channel boundaries or other vessels.

Improvement in Mathematical Models

The current practice for developing mathematical models for simulators is based on extrapolation of hydrodynamic coefficients from towing-tank tests for a restricted set of hull shapes. This practice may result in a degradation of the validity of the information when applied to conditions of loading and trim or to ships that differ from the model tests. In addition, the simulation of towed vessels is severely limited by the absence of systematic test data.

The conduct of full-scale real-ship experiments would significantly advance the state of practice in the development of mathematical models. These experiments could supplement the limited information available for shallow and restricted water, slow speed, and reverse propeller operational information. Vessels currently part of the U.S. Maritime Administration's Ready Reserve Fleet and some vessels in the Navy's Military Sealift Command fleet represent a possible resource for data to validate and improve mathematical models. In general, computational methods for determining the pertinent hydrodynamic parameters based on theories offer the possibility of more general and accurate simulations, particularly for ship operations in restricted waters and ship-to-ship interactions.

Recommendation 21: The U.S. Department of Transportation should develop standards for the simulation of ship maneuvering. The fidelity of the models should be validated through a structured, objective process. Standard models should be selected and tested in towing tanks and the results compared to selected full-scale real-ship trials of the same ships to provide benchmark data for validation and testing of simulators.

Recommendation 22: The U.S. Department of Transportation should initiate research to integrate computational hydrodynamics analysis with simulators in real time.

Funding Simulator-Based Training and Licensing

Specialized training on manned-model and computer-based simulators is not affordable to most individual mariners. The improvements in mariner competence and professional development possible through the application of simulator-based training are discussed throughout this report. Professional development is a shared responsibility among mariners, shipping companies, unions, port authorities and facility operators, and others. Each of these groups, as well as the general public, benefits from improved mariner competency and safety.

It is important, therefore, that any decision to mandate training for licensing or renewal include full consideration of options and mechanisms available to ensure that the training is available and affordable to all effected mariners.

Recommendation 23: The U.S. Coast Guard and U.S. Maritime Administration should assess the options for funding simulator-based training and licensing.

1

Mariners—Their Work and Professional Development

Recently, the professional competence of mariners has come under intense public scrutiny. Marine accidents, such as the grounding of the *Exxon Valdez* in Prince William Sound, Alaska, in 1989, have resulted in extensive marine pollution and damage to the marine environment and, in some cases, substantial loss of life. These accidents have involved U.S.-flag and foreign-flag seagoing ships and coastwise and inland tugs with barges. The impact and public perception of these marine disasters have created a need for a better understanding of the causes.

Casualty investigations show that, in many cases, accidents have occurred on well-equipped and seaworthy ships, with competent crews, often with a pilot on board, and in reasonable weather. A number of recent National Transportation Safety Board (NTSB) investigations have found that the primary cause of many marine accidents was an error of omission on the part of the deck officer that was either undetected or not articulated by others on the ship until it was too late.

Such human errors are exacerbated by weaknesses in traditional ship-bridge organization and management. For groundings, statistics suggest that as many as 71 percent are caused by bridge management error (Wahren, 1993). The well-publicized grounding of the *Queen Elizabeth 2* in 1992 is an example. Among its findings, the NTSB enquiry (1993a) criticized the failure of bridge management and teamwork and indicated that this failure was a major contributing cause of the accident.

> The National Transportation Safety Board determines that the probable cause of the grounding of the *Queen Elizabeth 2* was the failure by the pilot, master, and watch officers to discuss and agree on a navigation plan for departing Vineyard Sound and to maintain situational awareness after an unplanned course change.

Contributing to the accident was the lack of adequate information aboard the *Queen Elizabeth 2* about how speed and water depth affected the ship's under-keel clearance.

There are many other examples of human errors:

... the probable cause of the *Ziemia Bialostocka's* ramming of the Sidney Lanier Bridge was a failure of the pilot to maneuver the vessel properly because he did not make himself aware of and use all available maneuvering information and his failure to stop the vessel when he realized that it was not responding as he expected (NTSB, 1988b).

... the probable cause of the grounding of the U.S. tank ship *Star Connecticut* was the failure by the *Star Connecticut's* master and the Hawaiian Independent Refinery's mooring master to plan and coordinate the vessel's departure from the single point mooring buoy, which resulted in the master's inability to focus on and prioritize the critical tasks associated with departing the single point mooring buoy while maneuvering close to a shoal area known to have unpredictable ocean currents (NTSB, 1992).

... the probable cause of the collision between the *Juraj Dalmatinac* and the *Fremont* tow was the failure by the ship's pilots and master to effectively use all available equipment and personnel to evaluate the developing situation so that they could take timely action to avoid the collision (NTSB, 1993b)

... the probable cause of the ramming of the *Mont Fort* by the *Maersk Neptune* was the failure of the pilot to use the information concerning the radar distance to the anchored vessel provided by the master and the use of excessive speed while approaching his intended anchoring location (NTSB, 1988a).

Merely providing more equipment will not prevent accidents. Improved training and education can help prevent maritime accidents and casualties, and many agencies and companies are looking to simulation as one vehicle for improving mariner competency.

To make effective use of simulation it is necessary to assess both its benefits and limitations. This knowledge can then be used to develop a technical basis for informed national and international decision making about the appropriate role for simulation in the professional development and qualification of mariners.

THE MARINER POPULATION

Understanding the mariners who form the trainee population—their professional development needs and how these needs are met—is fundamental to determining the suitability of marine simulation for professional development and licensing.

The seagoing workforce is traditionally divided into three broad groups—deck, engine, and steward. Deck personnel, who are the focus of this report, can be divided loosely into two groups—licensed and unlicensed. The *licensed*

> **BOX 1-1**
> **Selected Definitions**
>
> **Bridge resource management** is a means of organizing the maritime bridge watch to use all resources effectively, including elimination of traditional vertical and horizontal barriers to coordination, communication, and integration.
>
> **Bridge team management** training includes passage planning and team management training that provides participants an opportunity to analyze various navigation scenarios and to demonstrate organizational procedures to assist in the safe conduct of the ship.
>
> An **integrated bridge** has centralized ship control and navigation systems. Navigational, control, and environmental data from differing sources are collected, processed, analyzed, and displayed to facilitate the navigation and control of the ship.
>
> **Marine certification and marine licensing** include the qualification requirements for the issuance of marine certificates of competency (e.g., marine licenses), the process by which these qualifications are fulfilled, examinations of knowledge and demonstrations of competence, administration of licensing programs by cognizant marine licensing authorities, and, by implication, official accountability for performance. For this report, the term *marine certification* is applied to the international process, and *marine licensing* is applied to the U.S. Coast Guard program.
>
> **Navigation** is the process of directing the movement of a ship from one point to another.
>
> A **pilot**, also referred to as a marine or maritime pilot, is an individual who operates from an organized pilot association or group and is licensed by a government authority (federal, state, or local authority empowered by a state) to provide pilotage services over specific waters or routes.
>
> **Piloting** refers to specialized work done by pilots. It is the act of directing and controlling the navigation and maneuvering of a vessel in pilotage waters.
>
> **Shiphandling** is very special, close-quarters work done primarily by pilots. It is the control and navigation of a ship by use of engines, rudder, thrusters, and tugs, as needed, taking into account environmental factors such as tide, current, wind, and weather.
>
> **Ship maneuvering** is changes in course in open water, usually to avoid other ship traffic.

mariner group consists of two major subgroups—*deck officers*, who are a part of a vessel's complement, and *marine pilots* (see Box 1-1, Selected Definitions), who are independent of the vessel. Deck officers can be further subdivided into masters, chief mates, mates, and individuals in these three subgroups with federal pilot endorsements on their U.S. Coast Guard (USCG) license.

Unlicensed seamen are generally *able-bodied seamen*, some of whom may be working their way up to licensed status through personal study, sea service, on-the-job training, and, in some cases, commercial or union-provided courses. One additional category, *trainees*, consists of *cadets* who are working toward their first license within a structured academic program, usually followed by semistructured professional development consisting of on-the-job learning with some structured training.

The skills requirements, duties, and responsibilities of each of these mariner categories must be understood when establishing professional development and qualification requirements. Unlike most professions, where specialization is the norm, mariners are generalists, expected to be individually proficient and able to work effectively in the unique shipboard operating and business environment. The licensed deck officer's duties go far beyond the well-known responsibilities of navigator and watchstander, especially aboard modern, highly automated ships. These ships often have short turn-around times in port and are manned by smaller crew sizes who have specialized technical training and are expected to perform a wide range of tasks with minimum assistance.

Furthermore, the marine operating environment and associated operational risks are highly situational (NRC, 1994). Thus, the maritime profession and other trades and professions are not analogous. The unusual nature of shipboard work is illustrated by the following observations:

- The command structure aboard ships follows the traditional naval hierarchy.
- Deck officers need to have directive control because the ship is a self-contained unit that operates independently.
- Deck officers must be generalists because they are responsible for all ship operations and business decisions.
- Deck officers must be able to make timely decisions to respond to sometimes rapidly changing operating conditions and emergency situations.
- Masters have ultimate responsibility. There are no committees aboard ship to share responsibilities for vessel operations or fault for the consequences of any errors in judgment.
- Deck officers work with little or no assistance and only limited supervision during watches when they are directly responsible for the safe navigation of the vessel.
- A deck officer's authority must be immediate and absolute while on watch, subject to the command authority of the vessel's master.
- Deck officers must be capable of responding to all medical and shipboard emergencies.
- Unlike the practice for entry positions in most industries, young officers, even those on their first voyage, may assume unsupervised watchstanding responsibility.

The International Maritime Organization (IMO), through its Standards for Training, Certification, and Watchkeeping (STCW) guidelines (IMO, 1993) and emergency operations, provides an international baseline for the professional qualification of deck officers and other bridge personnel for seagoing merchant marine service. Compliance with STCW guidelines is mandatory for IMO member countries that have ratified these international requirements. However, provisions of the STCW guidelines have not kept pace with developments in marine operating capabilities and practices. The guidelines are currently undergoing extensive revisions, which will include guidance for the use of marine simulation (IMO News, 1994). The IMO's role in developing these standards and the provisions of the STCW guidelines are discussed in Appendix B.

Most pilots serve local ports and waterways and normally remain within a single pilotage jurisdiction in the country where they are licensed. Therefore, the IMO has not promulgated mandatory standards for marine pilots. It has, however, promulgated recommended guidelines for the administration of pilotage systems and for knowledge required of marine pilots (IMO, 1981).

Within most countries, the responsible national licensing authorities set independent, usually much more stringent, standards. The USCG is the responsible licensing authority in the United States for deck officers.

Deck Officers

Traditional Marine Disciplines

The deck officer must first be trained in the traditional marine disciplines. The IMO mandatory knowledge categories for masters and chief mates on ships of 200 gross registered tons or more (IMO, 1993) include the following:

- navigation (Box 1-1) and position determination;
- watchkeeping;
- radar equipment;
- magnetic and gyro compasses;
- meteorology and oceanography;
- shiphandling and maneuvering (Box 1-1);
- ship stability, construction, and damage control;
- ship power plants;
- cargo handling and stowage;
- fire prevention and fire-fighting appliances;
- emergency procedures;
- medical care;
- maritime law;
- personnel management and training;
- radio communication and visual signaling;
- life saving; and
- search and rescue.

Ship's Business

Ocean shipping is a complex, multimillion dollar seagoing business. Aboard ship, deck officers spend a significant part of their work day managing routine business activities in cooperation with other ship's officers and shore management. Ship's business can include accounting and budgeting; payroll calculations; labor contract compliance and personnel relations; port and cargo documentation; requirements of international, admiralty, and business law, codes, and regulations for carriage of commodities by sea; ocean transportation regulations, customs, and immigration activities; vessel and cargo documentation; charter parties; stowage and inventory control; crew business organization and business meetings; and human relations skills inherent in organizing a vessel's operating teams and managing the social-living environment.

Cargo Handling

Deck officers are responsible for ship's operations at the terminal, for stowage planning, cargo characteristics, the techniques for stowage and carriage of dry and liquid cargos, and the proper response to problems that arise in the course of carrying routine and hazardous cargos.

Emergency Preparedness

Ship's officers organize and train the ship's crews in emergency response. They are the vessel's emergency medical technicians and must be trained to dispense medications and drugs and provide first aid and medical attention. They respond to emergencies and control of damage resulting from collision, grounding, fire, or the escape of hazardous and contaminating cargos into the environment. There are no fire, medical, or police departments available to assist the mariner at sea.

Marine Pilots

Marine pilots and deck officers have many overlapping skills, but their duties and responsibilities are distinctly different; whereas ship's officers are generalists, pilots specialize in a narrower range of skills. Pilots are, however, expected to be proficient in many of the same operating and emergency skills as ship's officers, especially navigation, operations, local knowledge, and emergency procedures (NRC, 1994). Pilot knowledge and skills recommended by the IMO are summarized below and discussed in Appendix B.

In the United States, both national (federal) and port-specific (state) pilotage systems are in use. The federal system primarily involves:

- deck officers who have been certificated by the USCG to pilot U.S.-flag vessels over specific coastwise routes, and

- individuals who are not members of a vessel's crew, but who are similarly USCG-certificated and provide coastwise pilotage services, primarily to U.S.-flag ships.

State pilotage systems involve marine pilots, not members of a vessel's crew, who provide pilotage services to foreign-flag ships and U.S.-flag ships engaged in domestic and foreign trade.

Operating knowledge and skills required of pilots include a broad range of traditional marine disciplines (IMO, 1981; NRC, 1994):

- terrestrial and electronic navigation and piloting (Box 1-1);
- seamanship, including use of anchors;
- shiphandling;
- use of radar and automatic radar plotting aids;
- use of tugs for towing and maneuvering;
- meteorology, plus knowledge of local winds and weather;
- passage planning;
- nautical rules of the road and local navigation rules;
- operation of bridge equipment; and
- vessel traffic systems.

Local Knowledge

Pilots are experts in knowledge about specific local ports or bodies of water. They serve aboard a mix of vessel types with varying maneuvering characteristics and are experts in how a vessel handles in shallow and restricted waters. A pilot has an intimate knowledge of local waters, including aids to navigation, depths, channels, currents, local wind and weather conditions, courses and distances to steer from sea buoy to berth, and bottom topography. The pilot must be familiar with local customs and working practices of the port, terminal operations, local navigation rules and regulations, and port equipment.

Business Practices

In addition to providing pilotage services, a pilot is often called on by a ship's master for insight on business practices and local regulations within the port. Therefore, pilots must become knowledgeable about local customs and immigration regulations, maritime law, ocean transportation regulations, and local business practices relating to port operations and cargo movements.

Emergency Procedures

Pilots must be able to respond to emergency situations, such as loss of steering or propulsion, and must be knowledgeable in port procedures for control of damage resulting from collision, grounding, fire, or the escape of hazardous cargos into the environment.

U.S. AND INTERNATIONAL OPERATING ENVIRONMENTS

U.S. Merchant Marine Shipping

The status of the U.S.-flag merchant marine fleet has changed substantially since the end of World War II, from a position of dominance to that of a minor player. As documented in Table 1-1, the U.S.-flag fleet in 1955 accounted for more than 21 percent of all ships worldwide and 27 percent of the carrying capacity of ships over 1,000 gross tons. That dominant position has steadily declined to less than 3 percent of all ships and 4 percent of carrying capacity worldwide in 1992—and the decline continues. Today U.S.-flag ships play only a minor role in world shipping, including domestic ports and coastal waters.

While the U.S.-flag share of the world fleet dropped dramatically, the total capacity of the world fleet grew by a factor of five. From 1955 to 1992, the size of the average ship tripled. "Super ships," very large crude carriers in excess of 350,000 deadweight tons, joined the world fleet. From these numbers, two points are clear:

- U.S.-flag ships with U.S.-citizen crews are a small part of the world fleet.
- Deck officers' responsibilities have steadily increased with increasing ship size.

Paralleling the growth of super ships has been a decline in crew size due to the advent of automation and other technological improvements (see discussion below). The number of ships operating from 1955 to 1992 has increased by 58 percent, leading to greater channel and harbor congestion. These factors contribute to a significant increase in watch officer responsibility and a need for comprehensive professional development and high performance standards.

Smaller Crew Size

Financial and operational considerations have resulted in an international trend toward smaller, multidisciplined crews with a higher level of technical skills and broader deck officer responsibility. Because of the widely varied size and quality of crews aboard ships of different registry, marine pilots in particular must be capable of adjusting to different operating conditions, ranging from ships where traditional operating methods prevail to those where the latest, most technically sophisticated systems are used. Generally, U.S. pilots have shown that they are able to function in these widely differing conditions, although many of these situations require an increased emphasis on voyage planning and on full communications, understanding, and agreement among the pilot and deck officers in advance of pilotage transits (NRC, 1994).

TABLE 1-1 Historical Inventory of World and U.S.-Flag Ocean Ships Over 1,000 Gross Tons

Year	Flag	Number of Ships	Total Gross Tons (1,000s)	Total Deadweight Tons (dwt) (1,000s)	Average dwt	Percent Total dwt	Percent Number of Ships
1955	World	15,148	92,944	129,975	8,580		
	U.S.	3,304	25,250	35,539	10,756	27.34	21.81
1960	World	17,317	122,027	171,890	9,926		
	U.S.	2,929	23,754	32,568	11,131	18.95	16.90
1965	World	18,329	151,868	217,229	11,852		
	U.S.	2,376	20,684	28,283	11,904	13.02	12.96
1970	World	19,980	211,401	326,999	16,366		
	U.S.	1,579	15,529	21,346	13,519	6.53	7.90
1975	World	22,872	33,042	556,572	24,334		
	U.S.	857	12,301	17,694	20,646	3.18	3.75
1980	World	24,867	385,711	654,909	26,336		
	U.S.	864	16,020	24,090	27,882	3.68	3.47
1985	World	25,424	391,979	656,422	25,819		
	U.S.	744	15,444	23,847	32,052	3.63	2.93
1990	World	23,596	386,736	637,493	27,017		
	U.S.	636	16,103	24,342	38,274	3.82	2.70
1992	World	23,943	397,225	652,025	27,232		
	U.S.	619	15,466	23,254	37,567	3.57	2.59

NOTES: 1992 compared with 1955:
- 81 percent decrease in number of U.S.-flag ships.
- 58 percent increase in number of ships in the world fleet.
- 34 percent decrease in U.S.-flag deadweight.
- 402 percent increase in total world deadweight.
- 249 percent increase in U.S.-flag ship average deadweight.
- 217 percent increase in world ship average deadweight.

1992 compared with 1980:
- 28 percent decrease in number of U.S.-flag ships.
- 4 percent decrease in number of ships in the world fleet.
- 3 percent decrease in U.S.-flag deadweight.
- 0.4 percent increase in total world deadweight.
- 35 percent increase in U.S.-flag ship average deadweight.
- 3 percent increase in world ship average deadweight.

SOURCE: U.S. Maritime Administration (1955, 1960, 1965, 1970, 1975, 1980, 1985, 1990, 1992).

Crew reductions may marginally increase the pilot's workload and, at times, require the pilot to take a more active role in ship-bridge operations. In practice, however, the pilot is more often affected by the trend toward use of less-experienced crews with, at times, only limited proficiency in English than by any reduction in the size of the ship's crew (NRC, 1990).

Crew Rotation Practices

Except in a few cases, contractual hiring practices with U.S. maritime unions in the past two decades typically resulted in "rotating" rather than "permanent" assignments for watch officers, with the change-over often measured in weeks or months. As a consequence, watch officers were often less familiar with the equipment and procedures of a particular ship than personnel assigned permanently to the same vessel, a situation that could impair a watch officer's effectiveness. Recently, some U.S. shipping companies and maritime unions have been working to revise assignment practices so that the masters, chief mates, and, in some cases, second mates are permanently assigned to the same vessel.

Integrated Bridge Systems

At sea, automated ships have generally made the crew's job easier. This trend toward automation, however, has sometimes increased the complexity of the pilot's work. In particular, ships with integrated bridges (see Box 1-1) may have fewer officers and crew aboard to assist pilots than ships with traditional bridge configurations (NRC, 1994).

MARINER TRAINING, LICENSING, AND PROFESSIONAL DEVELOPMENT

Professional development of mariners is a shared responsibility of the mariner, maritime academies and training institutions, unions, operating companies, pilot associations, marine licensing authorities, and the IMO (see Box 1-2). The process by which mariner competence is maintained and overseen includes:

- professional development practices by individual mariners and the organizations with which they are affiliated or employed,
- marine licensing by regulatory authorities, and
- the establishment of accountability for performance through professional and official discipline.

The Professional Development Process

Mariner training basically takes place along two different paths, each serving a specific purpose in the initial training and continuing professional development process. For most mariners, the educational process is clearly task-oriented. It has been developed in response to the needs of the industry and to a long-standing tradition of service to the ship and its cargo that still dominates the seafarer's attitude toward the job and education. Training programs vary in purpose and population and include a full range of programs, from subject-specific refresher courses to complete programs leading to a marine license.

> **BOX 1-2**
> **Mariner Professional Development: Training and Licensing**
>
> **U.S. Maritime Administration (MarAd).** Supports the training of merchant mariners by approving course curricula for federal and state maritime academies. Operates the U.S. Merchant Marine Academy and provides partial funding support for state marine academies, including the acquisition, maintenance, and operation of training ships and simulators.
>
> **U.S. Coast Guard (USCG).** Responsible for merchant marine licensing and the licensing of federal pilots. Licensing program relies on a three-pronged approach consisting of (1) a stated minimal amount of sea service (now coupled with some equivalency credits for certain simulator-based training), (2) the recommendation of applicant's peers, and (3) a written examination testing the applicant's theoretical knowledge. The USCG approves all courses that are prerequisites for marine licenses, including simulation-based courses. The USCG consults with the MarAd with respect to curricula used at the federal and state maritime academies.
>
> **State Pilotage Authorities.** Primarily responsible for official oversight for qualifying and licensing individuals who provide pilotage services to foreign-flag and U.S.-flag ships engaged in international trade. Pilotage systems at the state level vary in their content and administration (NRC, 1994).
>
> **International Maritime Organization (IMO).** Coordinates and promulgates the development of international guidelines and standards for training, certification, and watchkeeping for mariners engaged in ocean and near-ocean voyages. Develops and recommends guidelines for maritime pilots.
>
> **Labor Unions.** Active in training; some operate substantial training facilities.
>
> **Operating Companies (shipping and towing).** Involvement in training programs is determined by company policy.

Traditionally, mariners have prepared to qualify for their first marine license (third mate) through two basic professional development paths:

- structured marine education (at the baccalaureate level in the United States), which is the route followed by cadets to the third mate's license, and
- learning by experience (i.e., on-the-job training), which is the route generally followed by the able-bodied seaman to the third mate's license.

Once an individual has received the third mate's license there are no formal continuing-education requirements. However, after receiving that license most mariners pursue advancement through a combination of progressive education and sea service.

The first of these approaches—the structured approach—is the predominant route for initial training and licensing. This approach includes a comprehensive education program that satisfies minimum sea-service requirements, while exposing the prospective third mate to the full range of technical knowledge needed to serve as officer in charge of a navigational watch and as master.

Learning by experience, the second path, involves service aboard ship or other vessels. This experience is sometimes supplemented by short training courses at union training facilities or specialized license-preparation courses, such as those given at commercial license-preparation schools.

Advancement from the third mate to the master's license usually involves layered periods of classroom training and sea service.

Both approaches are in use throughout the world. Each has advantages and disadvantages. In the United States, the structured education approach prevails for deck officers, while on-the-job learning dominates in the towing industry. Learning by experience also prevails in the piloting profession, although many marine pilots are recruited from the pool of licensed deck officers in the shipping and towing industries. This practice can result in a beneficial combination of formal education and extensive practical experience (NRC, 1994).

The following sections summarize the primary sources of mariner professional training and development and their sources of support, where applicable.

Maritime Academies

U.S. Merchant Marine Academy. The U.S. Merchant Marine Academy at Kings Point, New York, is supported by the federal government through appropriations administered by the U.S. Maritime Administration (MarAd). The Academy offers deck and engine cadets a broad-based curriculum stressing traditional marine skills, with an emphasis on technology and maritime business, leading to a baccalaureate-level degree. Its students live in a military environment that prepares them for both their obligations as U.S. Naval Reserve officers and the disciplined merchant vessels shipboard environment. The Academy primarily trains entry-level officers, although the curriculum, by design, prepares cadets for all levels of responsibility. Graduates are prepared for the immediate responsibilities of a third mate or third assistant engineer and a U.S. Naval Reserve junior officer with the rank of ensign.

The mission of the Academy's continuing-education program is to provide specialized courses, seminars, and training programs for the U.S. maritime community and for government, commercial, and military personnel. These programs are primarily directed at increasing operational efficiency, improving safety standards, reducing accidents, providing technical and management skills, and reducing the possibility of environmental incidents.

State Maritime Academies. State maritime academies, located in California, Maine, Massachusetts, Michigan, New York, and Texas, offer similar, degree-granting programs to those of the U.S. Merchant Marine Academy. The state academies receive some government appropriations, including provision of federally owned school ships and their annual drydocking repairs. Federal funds are also provided for training equipment, such as computer-based marine simulators. There are differences in training programs and graduate military and service obligations among the state and regional academies and the federal maritime academy, but these differences are not pertinent to this report.

Labor-Management or Union Programs

Labor-management (i.e., union) training schools primarily provide technical training to upgrade skills for advancement and continuing professional education. These schools offer a wide range of professional training, including five schools that offer computer-based ship-bridge simulation training. Many of these schools' programs are customized to meet immediate needs of a student group or contract company. Mariners typically complete this training to qualify for a higher license, prepare for assignment on a new or different type of ship, or as nonmandatory training in fields such as bridge team or bridge resource management (see Box 1-1).

The costs of operating labor-management schools are generally included in collective bargaining agreements. Training is often at little or no direct cost to the individual mariner, since the expense of operating the schools comes from the negotiated training funds (though the mariner pays indirectly through deductions that might otherwise be part of an officer's wage package).

The shrinking U.S. merchant marine fleet and shrinking employer pool in the seagoing maritime industry has caused some union schools to open their facilities to other maritime and nonmaritime contract training. At various times, some schools have received federal support to acquire equipment and conduct training, such as mariner preparation to support military-force deployments during Operation Desert Storm.

Ship Owners and Operators

Vessel owners and operating companies commonly offer some type of shipboard training in subjects ranging from emergency procedures to operation and maintenance of new and existing shipboard equipment. Many operators also provide continuing education and training programs through contracts with commercial and union facilities. The quality of training and the extent to which ship owners train their officers vary considerably among companies, depending on the company's size, its attitude toward its shipboard staff, and the degree to which the company takes responsibility for its officers' performance.

The maritime industry is distinguished by the small size of individual companies and the limited number of officers employed. In many cases, these factors limit the amount of training done at the company level. There is no statutory obligation for employers to train their officers, other than some limited responsibility under the doctrine of seaworthiness. These conditions make it necessary to look beyond employers to consolidated programs operated by the maritime academies, labor-management schools, and private contractors to provide continuing education and training for mariners.

Vessel operators pay the cost of employer-provided, school-based, and shipboard training. A highly competitive market and a small employer base may make it difficult for some vessel operators to finance mandated training or extensive formal continuing-education programs. Operators have been and will continue to be an effective funding source for training directly related to their specific operations.

Traditional Colleague-Assisted Training

Colleague training aboard ship remains an important source of deck officer training. Historically, experienced officers have taken an active role in improving the skills of newer officers and crew members. Experienced officers have passed professional standards and procedures from generation to generation through formal and informal training. The extent and effectiveness of this training obviously varies greatly from ship to ship, depending on the aptitude for teaching and the attitude of the master and senior officers toward training. Given the minimal requirements for other forms of continuing education, a good case can be made that peer education is ultimately one of the most important sources of training. Unfortunately, this training, while important, is informal and may be inconsistent, and it does not meet all requirements for meaningful, continued professional development.

Pilot Apprentice Programs

In contrast to third mates, marine pilots are only given significant responsibility after completing an extensive apprenticeship. In most cases, these programs are conducted by local pilot associations, following traditional pilot development practices. Apprentice pilots make repeated passages over routes served by the association under the supervision of an experienced pilot (NRC, 1994).

Historically, U.S.-port pilot associations have maintained specialized training programs where apprentice pilots spend extensive time—commonly two to four years or longer—training aboard ship under the tutelage of senior pilots. These programs provide apprentice pilots experience aboard a wide variety of ships under a broad range of operating conditions before being placed in a position of full responsibility.

Several experts who met with the committee reported that, in some ports, pilot associations are adding requirements that apprentices be maritime academy graduates. These experts stated that a growing number of associations supplement hands-on training with in-house classroom training or with contract classroom and simulator-based training in radar and automatic radar plotting aid operations, bridge team or bridge resource management, shiphandling, and emergency procedures.

In some cases, pilots receive subsequent training for new and specialized ship types or for developing shiphandling techniques in modified ports and channels. The trend toward extended training programs and continuing professional development is expected to continue to respond to public expectations for marine safety, to upgrade pilot services for larger, more specialized ships, and to prepare for automated ships with smaller crew sizes (APA, 1993; NRC, 1994).

The cost of pilot training, which is currently a concern (especially for small-port pilot associations), is covered, if at all, as part of local pilotage rate structures. This cost varies, based on the pilot's professional background and training needs. Development of pilots from candidates with no prior maritime experience continues in several major pilotage associations. This practice requires a commitment of substantially greater resources than development of individuals with prior seagoing service or maritime education. To the extent that pilot candidates are recruited from the ranks of experienced mariners, including graduates of one of the maritime academies, the federal or state government will have contributed to the individual's professional development. To a limited extent, professional development costs may be subsidized by union and professional organizations.

Continuing professional development using advanced training media, especially marine simulation, though long used by several pilot associations, is relatively new in many pilotage systems. Given the organization and structure of U.S. pilotage, the costs of these programs must be borne within each local pilotage system. Some ports may not be capable of providing a sufficient funding base to support the cost of simulator-based training.

Other Organized Training Sources

Training programs are also available from private simulator-based training contractors and specialized educational facilities (see Box 2-2), such as the Seaman's Church Institute in New York, a charitable organization supported by donations. The Seaman's Church Institute provides fee-based and subsidized training, depending on need. The programs include a wide range of computer-based ship-bridge simulation training similar to that provided by commercial training facilities.

Individual Mariner-Funded Training

On a limited basis, individual mariners may also pay for training, such as short courses at license-preparation schools that provide training directly related to licensing examinations and renewals. Individuals also sometimes pay the cost of other, very limited training, primarily to prepare for entry-level positions in small commercial vessel operations. For practical reasons, the high cost of advanced training—particularly simulator-based training—is rarely paid by the individual mariner. It seems unrealistic, therefore, to expect any new or expanded training to be funded by the individual mariner, with the exception, perhaps, of some marine pilots who, working as independent contractors, would consider the cost of training to be a cost of doing business.

Marine Licensing

In its most basic form, a marine license is a document stating that the individual to whom it is issued meets the regulatory requirements (experience and knowledge) for the stated capability of the license (e.g., vessel categories and size and level of qualification—master, chief mate, second mate, third mate, pilot, operator of uninspected towing vessel). The license serves as an authorization allowing the holder to serve as a member of a vessel's complement or a pilot.

In general, a marine licensing authority's official responsibility is to regulate, according to established criteria, the license holder's authority to serve aboard vessels. A mariner's service is also professionally regulated, to varying degrees, by union and operating company policies and practices, or, in the case of independent marine pilots, by pilot associations. (The terms marine certification and marine licensing are defined in Box 1-1.)

Licensing authorities usually do not implement or operate professional development programs, although licensing requirements necessarily strongly influence the nature and form of professional development programs. To the extent that marine licensing *requires* or *permits* courses or programs, licensing authorities have domain over accreditation of educational and training curricula, programs, facilities, testing media (such as marine simulations), instructors, evaluators, and license assessors. In some cases, certain courses are required. For example, mariners of all flag states that have ratified the STCW guidelines must usually complete a training course to receive radar observer certification.

Issuance of a license by a licensing authority implies competency based on that authority's established minimum competency thresholds. The license level and type issued to deck officers and pilots depends on the industry sector where mariners are employed and the level of responsibility for which they are being licensed.

License Standards for Deck Officers in the United States

Federal licensing examinations are the basic professional assessment for deck officers serving aboard U.S.-flag vessels. The examinations are administered by the USCG to applicants who certify that they have acquired the prerequisite sea service and other educational equivalents (see Appendix C). Currently, tests for original licenses are comprehensive, multi-part written examinations. Most sections of the examination are multiple-choice questions. Higher-level licenses may be obtained at any point in an officer's career, once he or she has the required period of seagoing experience and passes a written examination similar in form to the original test. At present, there is no requirement for licensed officers to have any specific level of formal education to qualify for a marine license.

Renewal of the deck officer's license is usually required at five-year intervals, the maximum interval permitted by the mandatory provisions of the STCW guidelines. If the mariner is actively working in the maritime industry, the USCG's renewal examination is relatively perfunctory. The individual's ability to perform effectively under that license is not tested during license renewal.

Changes in Deck Officer Licensing in the United States. USCG records for issuance and renewal of "any gross tons" or "unlimited" licenses for the years 1986 through 1993 are summarized in Table 1-2. These records show a reduction in the new licenses issued and a drop in renewals for all deck officer categories during those years. New issues are down 26.6 percent; renewals are down 14.7 percent. These declines are consistent with the steady decline of the size of U.S.-flag fleet and the gradual reduction of ship crew size, regardless of the increase in individual ship size.

Geographical Distribution of Deck Officer Licenses. The committee examined the regional distribution of relevant deck officer license activity and its relationship to the location of known ship-bridge simulator facilities. These data are summarized in Table 1-3. When the committee began its deliberations in 1992, there was some concern that the preponderance of simulator facilities were located on the East Coast, with a dearth of facilities in the West and Gulf coasts. From a review of the market in early 1995, however, it appears that the geographical gaps in simulator availability are being filled, and expansion is expected to continue as industry demand for simulator-based training services grows.

Table 1-4 shows total 1994 USCG activity for limited and unlimited licenses by category. The license activity data in Tables 1-3 and 1-4 are summarized from the USCG's (1994) unpublished report, *License Activity by Port*, which lists license activity at all USCG license-issuing locations. That report does not differentiate between "limited" and "unlimited" licenses. Consequently, license issuance and renewal numbers, particularly in the master's category, represented in Tables 1-3 and 1-4 are substantially greater than those for "ocean unlimited" in Table 1-2.

License Standards for Pilots in the United States

Pilot licenses are generally obtained through a two-track licensing examination process. Licensing is primarily administered by state pilotage authorities

TABLE 1-2 U.S. Coast Guard Ocean-Only[a] License Statistics for Deck Department, Any Gross Tons,[b] Fiscal Years 1986–1993

License Category	1986	1987	1988	1989	1990	1991	1992	1993	1986–1993 Total	1986–1993 Average per Year
Master										
New issues	173	181	128	120	113	120	166	115	1,116	140
Renewals	769	648	597	586	488	590	615	666	4,959	620
Total	942	829	725	706	601	710	781	781	6,075	759
Chief Mate										
New issues	129	161	77	69	69	113	85	118	821	103
Renewals	199	226	205	166	157	181	157	194	1,485	186
Total	328	387	282	235	226	294	242	312	2,306	288
Second Mate										
New issues	177	155	157	173	135	176	229	175	1,377	172
Renewals	275	247	216	199	141	174	184	214	1,650	206
Total	452	402	373	372	276	350	413	389	3,027	378
Third Mate										
New issues	511	420	315	297	224	467	378	319	2,931	366
Renewals	416	437	360	404	317	413	402	373	3,122	390
Total	927	857	675	701	541	880	780	692	6,053	757
Total	2,649	2,475	2,055	2,014	1,644	2,234	2,216	2,174	17,461	2,183

[a]Issues and renewals only. No endorsements or failures.
[b]Any gross tons—limited licenses not included. The USCG advises that the above totals may be somewhat inflated by "double counts." A USCG computer search to identify duplications produced a 9.62 percent double-count factor for deck officers.
SOURCE: USCG (1986–1993).

under established laws and regulations. In nearly every case, the pilot association, either officially or by custom, administers apprentice programs for the route where the apprentice expects to work. Apprentice pilots make repeated trips, several hundred in many cases, under the supervision of experienced pilots aboard ships traversing the route for which the license is sought. The experienced pilot monitors and evaluates the apprentice's progress and is responsible for the actual vessel piloting. After documenting extensive route experience, the apprentice pilot is usually assessed by the pilot association and examined by the licensing authority (NRC, 1994).

Because virtually all pilots are required by their associations or state pilotage authority to hold a federal, first-class pilot's license or endorsement, the USCG also examines most pilots. In addition, most pilots must satisfy basic license requirements for mariners, and many hold some form of marine license prior to becoming an apprentice (NRC, 1994). To be licensed, the applicant for a federal pilot's license

TABLE 1-3 U.S. Coast Guard Limited and Unlimited[a] License Activity and Number of Facilities with Category I and Category II Simulators[b]: Summarized by Region[c]

Port	Issue	Renewal	Total	Percent Total	Number of Facilities with Category I and II Simulators
East Coast, North	1,694	3,227	4,921	26	8
East Coast, South	947	2,082	3,029	17	1
Gulf Coast	987	2,355	3,342	20	1
West Coast, North	646	1,227	1,873	11	2
West Coast, South	618	1,133	1,751	10	1
Alaska	352	214	566	3	0
Hawaii	167	465	632	4	0
Great Lakes	519	650	1,169	7	2
Mid-Continent	156	303	459	2	1
Total	6,086	11,656	17,742	100	16

[a]Licenses issued for operation of vessels over 1,600 gross tons are "unlimited."
[b]Category I are full-mission simulators.
  Category II are multi-task simulators.
  See Box 2-1 for complete IMO definitions of simulator categories.
[c]Regions and ports defined:
  East Coast, North—Boston, Massachusetts; New York, New York; and Baltimore, Maryland.
  East Coast, South—Charleston, South Carolina; and Miami, Florida.
  Gulf Coast—New Orleans, Louisiana; and Houston, Texas.
  West Coast, North—Seattle, Washington; and Portland, Oregon.
  West Coast, South—San Francisco, Los Angeles, and Long Beach, California.
  Alaska–Anchorage and Juneau.
  Hawaii—Honolulu.
  Great Lakes—Toledo, Ohio.
  Mid-Continent—St. Louis, Missouri; and Memphis, Tennessee.
SOURCE: USCG (1994).

or endorsement must demonstrate route knowledge by drawing from memory detailed charts of the pilotage route, showing aids to navigation, geographic and hydrographic data, and other information. The remainder of the examination is similar to the deck officer's license examination, with more-detailed testing of subjects related to shiphandling and local knowledge.

After receiving an unrestricted license, pilots are rarely formally re-evaluated. Measures to detect degraded performance, before the performance becomes a causal factor in a marine accident, are informal and unevenly applied. There is, however, a growing trend among pilot associations and state pilotage authorities to require pilot participation in continuing professional development programs (NRC, 1994).

TABLE 1-4 U.S. Coast Guard Total Limited and Unlimited[a] Licenses, by Category, 1994

License Category	Issue	Renewal	Total	Percent Total
First-Class Pilot	115	982	1,097	7
Master	4,483	9,292	13,775	77
Chief Mate	138	204	342	2
Second Mate	143	224	367	2
Third Mate	437	419	856	4
Mate	770	535	1,305	8
Total	6,086	11,656	17,742	100

[a]Licenses issued for operation of vessels over 1,600 gross tons are "unlimited."
SOURCE: USCG (1994).

Other Tests of Mariner Abilities

Excluding license examinations, the mariner's abilities are currently tested during his or her career in a number of ways at varying intervals by several evaluating bodies. These tests include:

- traditional testing through written examination during the normal four-year course of instruction at the state and federal maritime academies;
- instruction and testing at training facilities and union schools;
- formal periodic job evaluations by employers, based on company performance standards for job retention and promotion;
- continuous, informal on-the-job performance evaluations by masters and senior officers aboard ship; and
- pilot association examinations and check-rides and state license examinations for pilots.

There are no post-graduate education requirements for deck officers and pilots working aboard ship, although some practicing mariners do have post-graduate degrees in business, law, or other subjects associated with seagoing work. Some mariners periodically take refresher and upgrade courses to stay current. In addition, all actively sailing deck officers must attend radar training courses prior to each license renewal.

Some employers require deck officers to take retraining and refresher courses applicable to their shipboard positions. Some companies perform annual performance evaluations to determine continued mariner competency and monitor employees' career progression. These evaluations are normally conducted by the mariner's senior officer or a company-designated official, usually a senior master. Because the aim is to enhance performance at all stages of the officer's career, evaluations are usually interactive between reviewer and officer. Some companies also include counseling as part of the evaluation process. Evaluations are reviewed by the employer's human resource staff ashore to determine promotion capabilities, and promotions and continued employment are often based on evaluation results.

The Need for Improved Professional Development Systems

Traditional approaches to professional development have produced highly competent mariners, although not always systematically and with varying individual qualifications. Marine accidents involve experienced and competent mariners. Furthermore, most investigations include the identification of an error chain involving those mariners. The implication is that needed additional or special professional development is not always provided through the traditional approach. Marine simulation has become a major focus as one method of addressing these concerns.

U.S. Mariner Professional Development

A growing number of shipping and towing companies, members of the piloting profession, marine educators, and worldwide marine licensing authorities, including the USCG, have recognized the need for improved professional development, licensing, and certification programs. Some operating companies and pilot organizations have seriously begun to review and revise training needs to improve professional performance. These groups have developed training programs intended to achieve the objective of improved professional competence and are increasingly using marine simulation (APA, 1993; Mercer, 1993; Muirhead, 1993; NRC, 1994).

There is, however, a diversity of opinion among experts about real training and licensing needs and the suitability of simulation as a medium for satisfying these needs. This diversity is fueled by the fact that, although many of the tasks and subtasks essential to safe and effective vessel operation have been well described (Hammell et al., 1980), much of the work is dated, and there has been little systematic effort to correlate task and subtask descriptions with the nature and form of training. Consequently, there is little reinforcement through practice and task replication of the knowledge, skills, and abilities needed during actual operations.

A comprehensive understanding of job tasks and performance is fundamental for many reasons, including:

- identifying professional development needs;
- developing and conducting effective educational, training, and licensing programs (including options for remission of sea time);
- determining the short-term and long-term effectiveness of professional development programs; and
- assessing human performance in the classroom and on the job.

> **BOX 1-3**
> **The National Vocational Qualification (NVQ)**
> **System of the United Kingdom**
>
> **Aims of the NVQ system:**
>
> - Improve vocational qualifications by basing them on the standards of competence required in employment.
> - Establish a simple framework that will facilitate (1) access, (2) progression, and (3) continued learning.
>
> **There are five levels of NVQ. Each level has a competence definition. A sample of the levels of interest to this study are as follows:**
>
> - Level 1—(in the marine field this might be descriptive of a seaman): competence in the performance of a range of varied work activities, most of which may be routine and predictable.
> - Level 3—(class 3 license): competence in a broad range of varied work activities performed in a wide variety of contexts, most of which are complex and nonroutine. There is considerable responsibility and autonomy, and control of guidance of others is often required.
> - Level 5—(master's license): competence that involves the application of a significant range of fundamental principles and complex techniques across a wide and often unpredictable variety of contexts. Substantial personal autonomy and often significant responsibility for the work of others and for the allocation of substantial resources feature strongly, as do personal accountability for analysis and diagnosis, design, planning, execution, and evaluation.
>
> **For each level (by vocation) there are detailed definitions of:**
>
> - basic knowledge required,
> - tasks and subtasks necessary to perform,
> - performance criteria, and
> - assessment methodologies.
>
> **For maritime operations, NVQ is investigating the use of simulators for assessment within the system.**
>
> SOURCE: Implementing NVQs Workshop (1994).

International Professional Development

The steady increase in foreign-flag ships in U.S. waters (and the steady decrease of U.S.-flag vessels) presents an additional mariner competency concern. U.S. influence on the training and certification of international mariners is primarily through the IMO and the STCW guidelines. Since ratifying the STCW guidelines in 1991, the United States has become a leader in efforts to upgrade and improve the guidelines by "placing emphasis on stricter regulatory control for international shipping, and incorporation of modern training and certification methods, including use of simulators" (Drown and Mercer, 1995). Without these improvements in international standards, there can be no meaningful reduction in accidents in U.S. waters.

The United Kingdom has undertaken a major effort to improve the professional competency of all major vocations, including mariners. Box 1-3 summarizes the objectives of the U.K.'s National Vocational Qualification system, a system being structured to define levels of advancement within each vocation and to identify training and assessment certification criteria at each level in each vocation.

The Need for a Systematic Approach to Professional Development

It has been a long-standing practice to focus on knowledge as the basic determinant of mariner competency. The almost exclusive emphasis on knowledge—and general absence of systematic attention to job tasks and performance in training and marine licensing ship's officers—results in heavy reliance on intuition in mariner professional development and qualification.

There are no professional standards for skills development to guide and optimize the content and emphasis of on-the-job training for mariners. The result is a fundamental weakness in the traditional approach to professional development and qualification, because the content and emphasis of training courses and license examinations may not correspond to actual needs. This situation is exacerbated by the relative lack of data to guide a more scientific assessment.

Marine simulation, applied in a structured program, based on relevant and focused-task and subtask analyses and skills correlation, represents significant potential to supplement traditional training and professional development (see Chapter 3). Integrating simulation into structured training can ensure high levels of competency nationally and internationally. All forms and levels of simulation could be effectively applied to enhance original learning and refresher training.

REFERENCES

APA (American Pilots' Association). 1993. APA promotes BRM [bridge resource management] training for pilots. Press release, American Pilots' Association, Washington, D.C., October 5.

Drown, D.F., and R.M. Mercer. 1995. Applying marine simulation to improve mariner professional development. Pp. 597–608 in Proceedings of Ports '95. New York: American Society of Civil Engineers.

Hammell, T.J., K.E. Williams, J.A. Grasso, and W. Evans. 1980. Simulators for Mariner Training and Licensing. Phase 1: The Role of Simulators in the Mariner Training and Licensing Process (2 volumes). Report Nos. CAORF 50-7810-01 and USCG-D-12-80. Kings Point, New York: Computer Aided Operations Research Facility, National Maritime Research Center.

IMO (International Maritime Organization). 1981. Training, Qualification and Operational Procedures for Maritime Pilots Other Than Deep-Sea Pilots. IMCO Resolution A.485(XII) adopted on November 19, 1981. London, England: IMO.
IMO (International Maritime Organization). 1993. STCW 1978: International Convention on Standards of Training, Certification, and Watchkeeping, 1978. London, England: IMO.
IMO News. 1994. World maritime day 1994: better standards, training, and certification—IMO's response to human error. IMO News (3):i-xii.
Implementing NVQs (National Vocational Qualifications) Workshop. 1994. Business and Technology Educational Council. United Kingdom. Issue 1. September.
Mercer, R.M. 1993. Research and Training Aspects of Ship Simulators from an Educational Perspective. Pp. 387–395 in MARSIM '93. International Conference on Maritime Simulation and Ship Maneuverability. St. Johns, Newfoundland, Canada, September 25–October 2.
Muirhead, P. 1993. Marine simulation performance measurement and assessment: methodologies and validation techniques: a critique. Pp. 417–426 in MARSIM '93. International Conference on Maritime Simulation and Ship Maneuverability. St. Johns, Newfoundland, Canada, September 25–October 2.
NRC (National Research Council). 1990. Crew Size and Maritime Safety. Committee on the Effect of Smaller Crews on Maritime Safety, Marine Board. Washington, D.C.: National Academy Press.
NRC (National Research Council). 1994. Minding the Helm: Marine Navigation and Piloting. Committee on Advances in Navigation and Piloting, Marine Board. Washington, D.C.: National Academy Press.
NTSB (National Transportation Safety Board). 1988a. Ramming of the Maltese Bulk Carrier *Mont Fort* by the British Tank Ship *Maersk Neptune* in upper New York Bay, February 15, 1988. Mar-88/09. Washington, D.C.: NTSB.
NTSB (National Transportation Safety Board). 1988b. Ramming of the Sidney Lanier Bridge by the Polish Bulk Carrier *Ziemia Bialostocka*, Brunswick, Georgia, May 3, 1987. Mar-88/03. Washington, D.C.: NTSB.
NTSB (National Transportation Safety Board). 1992. Grounding of the U.S. Tank Ship *Star Connecticut* in the Pacific Ocean, near Barber's Point, Hawaii, November 6, 1990. Mar-92/01. Washington, D.C.: NTSB.
NTSB (National Transportation Safety Board). 1993a. Grounding of the United Kingdom Passenger Vessel RMS *Queen Elizabeth 2* near Cutty Hunh Island, Vineyard Sound, Massachusetts, August 7, 1992. Mar-93/01. Washington, D.C.: NTSB.
NTSB (National Transportation Safety Board). 1993b. Collision of the U.S. Towboat *Fremont* and Tow with St. Vincent and the Grenadines Registered Containership *Juraj Dalmatinac*, Houston Ship Channel, December 21, 1992. Mar-93/02. Washington, D.C.: NTSB.
USCG (U.S. Coast Guard). 1986–1993. Marine Safety Council Proceedings Magazine. Washington, D.C.: USCG.
USCG (U.S. Coast Guard). 1994. Licensing Activity by Port. Unpublished report, USCG, Washington, D.C.
U.S. Maritime Administration. 1955, 1960, 1965, 1970, 1975, 1980, 1985, 1990, 1992. Annual Vessel Inventory Reports. Washington, D.C.: U.S. Department of Transportation.
Wahren, E. 1993. Application of airline crew management training in the maritime field. Pp. 591–599 in MARSIM '93. International Conference on Maritime Simulation and Ship Maneuverability. St. Johns, Newfoundland, Canada, September 26–October 3.

2

Use of Simulation in Training and Licensing: Current State of Practice

Because of recent trends in the marine industry toward smaller crew size, heightened public concern about marine safety and expectations for improvements, and changes in navigation and ship control technology, the integration of marine simulation into mariner training programs offers advantages and opportunities to improve human performance in a safe environment.

RATIONALE FOR USING SIMULATORS

In this report, the *simulator* refers to the hardware or apparatus that generates the simulation. *Simulation* refers to the representation of conditions approximating actual or operational conditions. Simulations can be formalized into scenarios that are used for teaching and performance evaluation. A *scenario* is a specific simulation with a specific objective.

The theoretical rationale for the use of simulators for training is based on the concept of skill transfer—that is, the ability to adapt skills learned in one context to performance or task execution in another. Because no situation is ever identical to a previous experience, the fact that an individual becomes more skilled with each repetition of a similar task attests to the fact of transfer. Indeed, a faith in the "fact" of transfer constitutes the basic justification for all formal training programs. It is assumed that skills and knowledge learned in a classroom can be applied effectively to relevant situations outside the classroom.

No training environment will be exactly the same as the operating situation. To ensure that all training goals are met, it may be appropriate to supplement the learning with apprenticeships or a similar formal mechanism to reinforce learning.

Traditional classroom teaching has for generations been an effective method for teaching theory. Teaching methods usually include the instructor lecturing to the class, with the possibility of use of an overhead projector, chalkboard, or sometimes a movie or video to amplify training objectives. In the traditional setting, the instructor is in direct control and may or may not invite questions and discussion.

With the addition of simulation to the course curriculum, the instructor can fill the gap between theory and application (MacElrevey, 1995). The instructor can create an interactive environment where instructor and students actively participate in a demonstration applying theory to the real world (see additional discussion in Chapter 3).

TYPES OF SIMULATORS

Marine Simulator Classification

The physical (including engineering and technical) environment in which transfer of learning occurs consists of hardware, software, and the resulting displays and physical settings or conditions simulated. The physical environment and capabilities vary substantially among marine simulators. Unlike the highly structured environment of commercial air carrier simulators, with its well-defined classifications, technical specifications, and standards, the marine industry is just now developing a standard terminology for describing simulators. Industrywide technical specifications or performance standards have yet to be adopted.

The simulator classification system proposed for adoption by the International Maritime Organization (IMO) (see Box 2-1) is used in this report for consistency with current international developments. Under this system, simulators fall into four major categories—full-mission, multi-task, limited-task, and special-task simulators (also referred to as desktop or PC simulators).

Currently, there is no plan to include *technical* specifications for simulators in the IMO's efforts to revise the international marine Standards for Training, Certification, and Watchkeeping (STCW) guidelines. The STCW guidelines are expected, however, to include simulator *performance* standards to guide the effective and uniform use of simulators for marine professional development and certification. These performance standards are expected to prescribe minimum criteria that must be met: for example, field-of-view requirements for different types of functions and tasks such as watchkeeping and shiphandling (IMO News, 1994; Muirhead, 1994).

Within the marine industry, the International Marine Simulator Forum, an organization of simulator facility operators and other interested parties, and the International Maritime Lecturers Association, an international professional organization of marine educators and trainers, have been working to develop *technical* standards for simulators that would complement and support the STCW guidelines.

> **BOX 2-1**
> **Marine Operations Bridge Simulators**
> **Classifications Proposed to International**
> **Maritime Organization (IMO)**
>
> Within the marine industry, the terminology used to describe or classify simulators varies greatly. The terminology used in this report has been proposed for adoption by the IMO.
>
> **Category I: Full mission.** Capable of simulating full visual navigation bridge operations, including capability for advanced maneuvering and pilotage training in restricted waterways.
>
> **Category II: Multi-task.** Capable of simulating full visual navigation bridge operations, as in Category I, but excluding the capability for advanced restricted-water maneuvering.
>
> **Category III: Limited task.** Capable of simulating, for example, an environment for limited (instrument or blind) navigation and collision avoidance.
>
> **Category IV: Special task.** Capable of simulating particular bridge instruments, or limited navigation maneuvering scenarios, but with the operator located outside the environment (e.g., a desktop simulator using computer graphics to simulate a bird's-eye view of the operating area).
>
> SOURCE: Drown and Lowry (1993).

Simulator and Simulation Validity

Simulators and simulations vary greatly among facilities. Any discussion of simulator and simulation standardization needs to include issues of validation and validity. *Validation* is the process of evaluating specified characteristics of a simulator or simulation against a set of predetermined criteria. Assessing simulator or simulation validity generally includes consideration of two components—fidelity and accuracy. *Fidelity* describes the *degree of realism* or similarity between the simulated situation and real operation. *Accuracy* describes the *degree of correctness* of the simulation, with a focus on ship trajectory and location of aids to navigation and other critical navigational cues. The issues of simulator and simulation performance, technical standards, and validation are discussed in more detail in Chapter 7 and in Appendix D.

Computer-Based and Physical Scale-Model Simulators

A wide range of simulator capabilities are in use for training worldwide. Marine simulator capabilities for channel design and mariner training developed along two parallel and complementary lines—*computer-based simulators* and *physical scale models.* Computer-driven ship-bridge simulators, which originated in the 1960s, are used at many locations in the United States and worldwide. These simulators range from "blind pilot" (radar only) pilothouse mockups, to full-bridge mockups with projection systems, to fully equipped bridges on motion platforms approaching the level of sophistication associated with commercial and military aviation visual cockpit simulators.

The first computer-based simulators were based on simplified mathematical models[1] for a ship's hydrodynamics. These early models were coupled with rudimentary bridge mockups controlled by computers. Simulator technology has evolved with improvements in computer hardware, along with increasing knowledge from naval architects of appropriate models for ship dynamics. Ship-bridge simulators have also benefited from advances in computer-generated imagery (CGI) technology.

Complementary developments to ship-bridge simulator capabilities have occurred with the use of physical scale models of ships, referred to as *manned models*. The use of manned models was initiated in France in 1966. Manned models in the form of scale models of ships are used primarily for shiphandling training. Radio-controlled scale models have also been used for shiphandling training, but only to a very limited extent. Although scale models have not been developed for training in either the coastwise or inland towing industries, where on-the-job training for shiphandling is common practice, they have been used extensively for channel design and developing maneuvering strategies in new and unusual situations.

Figure 2-1 is a simple schematic of the types of marine simulators. Box 2-2 lists the locations of category I and II simulator facilities in the United States. Rapid advances in microcomputer hardware and software programming capabilities have also increased the number of microcomputer desktop simulators now available.

Full-Mission and Multi-Task Ship-Bridge Simulators

The study of ship maneuvering originated from the need to design ships with maneuverability characteristics that either meet specific requirements (turning circle diameter or tactical diameter was an early specified requirement) or were reasonable for the ship's mission. As the mathematical theory and hydrodynamics of ship movements advanced, more accurate computer-driven mathematical models were developed to represent and predict ship trajectories.

---

[1] A mathematical model is a collection of mathematical equations capable of simulating a physical situation. Mathematical models in the context of ship simulation refer to the modeling of the trajectory of a ship as its controllers (e.g., rudder, propeller, bow thruster) are activated. Models range from simple to complex, depending on the modeling of the hydrodynamic effects and the mechanical relationships present.

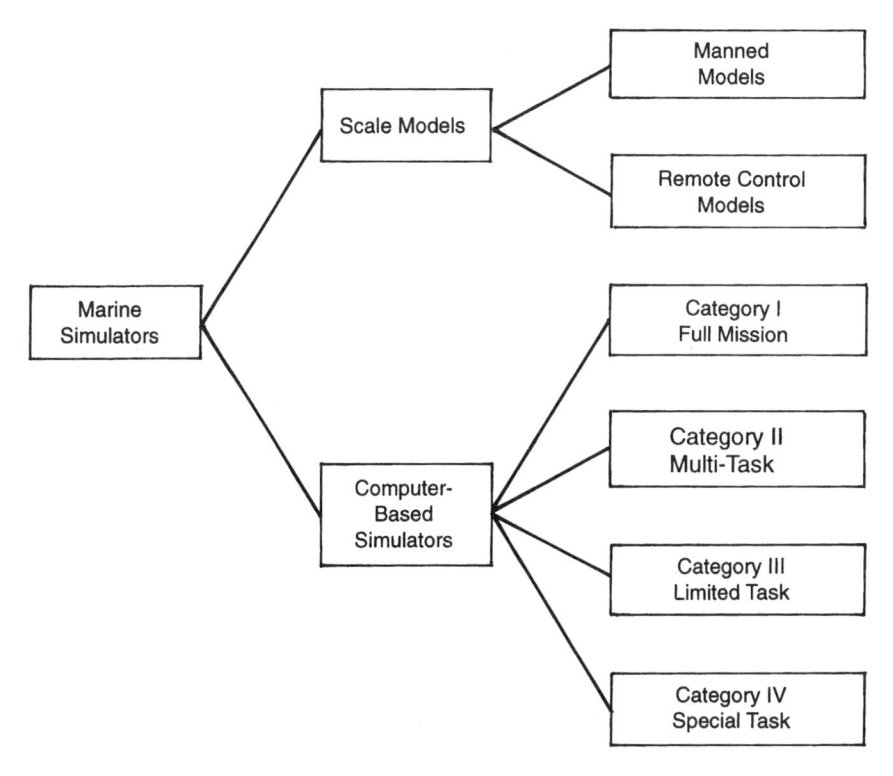

FIGURE 2-1 Types of marine simulators.

These early models used analog computers. With the introduction of digital computers, more complex models were developed for ship design. Ship-bridge simulators, capable of involving people in a real-time experience, were developed by combining digital computer-based models with bridge equipment, bridge mockups, and visual projection systems. As computer technology and CGI advanced, so did ship-bridge simulators. Modern computers made it practical to create ship-bridge simulators for full-mission and multi-task training. Computers also made it practical to combine actual radar equipment with mathematical models of vessel behavior to create radar simulators for use as an element of full-mission or multi-task training, or as a limited-task, stand-alone training device.

Full-mission, multi-task, and limited-task simulators are, as a rule, operated in real time and can appear to be quite realistic. This realism is referred to as *face validity* or *apparent validity* (NRC, 1992). Ship-bridge simulators are used for all types of operational scenarios. There are several important issues in using computer-based marine simulators (NRC, 1992), including:

- whether the appropriate vessel maneuverability cues are present and correctly portrayed in the simulation,

> **BOX 2-2**
> **Location of U.S. Facilities with
> Category I and Category II Simulators**[a]
>
> **East Coast, North Region**
> Maine Maritime Academy, Castine
> Massachusetts Maritime Academy, Buzzards Bay
> MarineSafety International, Newport, Rhode Island
> U.S. Merchant Marine Academy, Kings Point, New York
> SUNY Maritime, Bronx, New York
> Seaman's Church Institute, New York
> Maritime Institute of Technology and Graduate Studies, Linthicum Heights, Maryland
> Harry Lundeberg School of Seamanship, Piney Point, Maryland
>
> **East Coast, South Region**
> STAR Center, Dania, Florida
>
> **Great Lakes Region**
> Great Lakes Maritime Academy, Traverse City, Michigan
> STAR Center, Toledo, Ohio
>
> **Gulf Coast Region**
> Texas Maritime, Galveston
>
> **West Coast, North Region**
> STAR Center, Seattle, Washington
> California Maritime, San Francisco
>
> **West Coast, South Region**
> MarineSafety International, San Diego, California
>
> **Mid-Continent Region**
> U.S. Army Corps of Engineers, Vicksburg, Mississippi
>
> ---
>
> [a]In addition to the simulator facilities listed, the U.S. Naval Academy and the USCG Academy have category I and II simulators that are used exclusively by those institutions.

- whether the ship's maneuvering response is correct,[2] and
- the relative importance of accuracy in these areas.

Full-mission and multi-task ship-bridge simulators place the trainee inside a bridge mockup with actual bridge equipment or fully functional and configured

---

[2]The difficulties of properly modeling ship response are discussed in Appendix D, which deals with the complexities of hydrodynamic interactions, the nonlinear nature of ship motions, and their practical impact on trajectory responses in simulations.

emulations of bridge equipment. Figures 2-2 and 2-3 are views of the bridge of a full-mission simulator. The trainee is provided with correct or approximately correct angular relationships, depending on how projection screens are wrapped around the bridge. Normally, the complete ship-bridge simulator will integrate noise, normal information inputs and distractions from equipment, movement about the bridge, involvement of other bridge personnel, multiple tasks, and stress associated with the combined effects of these and other components.

In varying degrees, simulators take advantage of training, transfer, and retention benefits suggested by human performance literature for training systems approximating real operating conditions. From a technical perspective, in a high-fidelity, full-mission ship-bridge simulator, the training environment is expected to approach equivalency with the actual operating environment being simulated. The resulting simulation provides a relatively complete, realistic, and risk-free training environment suited for full-scale operational situations. Subject to the capabilities of individual simulators, the ship-bridge simulator-based training environment is suitable for such courses as:

- bridge team management;
- bridge resource management;
- shiphandling in open waters, channels, and waterways;
- docking and undocking evolutions (especially if equipped with bridge wings or configured as a bridge-wing simulator);
- bridge watchkeeping, including terrestrial and electronic navigation;
- rules of the road; and
- emergency procedures.

Despite current advances, simulators are not perfect. For example, limitations in the visual scene appear to encourage more reliance on electronic equipment, especially radar, than is common practice in unrestricted visibility conditions.

Limited-Task Simulators

Limited-task simulators place the trainee inside a training environment that is more limited in its capabilities to simulate navigation and collision avoidance situations. Limited-task simulators for underway vessel operations include radar simulators and "blind-pilot" simulators.

Radar Simulators. Radar simulators are an effective tool for mariner training. Developed independently from ship-bridge simulators, these simulators were first used for mariner training in the 1960s. Analog computers and coastline generators produced visual presentations on actual radar equipment. Simple linear equations of motion were used to predict ship trajectories.

Although digital coastline generators were available for military applications as early as 1973, transition to digital radar simulators in commercial marine applications followed the introduction of digital radars into commercial maritime

FIGURE 2-2 View of the bridge of a full-mission simulator. (STAR Center, Dania, Florida)

operations in the early 1980s. Today, virtually all radar simulators use digital data. The development of digital computers enables the use of sophisticated, mathematical trajectory and prediction models to drive radar simulators, either independently or as an element of a ship-bridge simulator. Some radar simulators are at the level of ship-bridge simulators, but without the visuals. Many are less sophisticated.

Radar simulators are operated in real time. When operated as stand-alone, radar simulators train mariners to use radar. The application of stand-alone radar simulators for ships' officers tends to focus on open-sea work and approaches to pilot boarding areas where shallow water and bank effects are either not present or not relevant to training objectives. For navigation and piloting training, where shallow or restricted water effects are often an element of the training objectives, it may be necessary to assess the capabilities of the equations of motion used to drive the radar simulator to determine whether radar simulation is suitable.

Because shallow water and bank effects were generally not considered important to training objectives, and because the computational capabilities for using sophisticated equations of motion were expensive, most earlier radar simulators featured lower-fidelity, less-sophisticated equations of motion than ship-bridge simulators. Thus, the representation of hydrodynamic interactions for waterway operations is, in general, less accurate than that achieved through use of latest-generation mathematical models. The latest generation of microcomputers

FIGURE 2-3 View of the bridge of a full-mission simulator. (STAR Center, Dania, Florida)

have the capability to use highest-fidelity equations of motion in stand-alone radar simulations; older microcomputers need to be linked.

"Blind-Pilot" Simulators. Blind-pilot simulators generally consist of a fully operating bridge mockup with a radar, but without visual CGI. Simulators of this type have been installed by a number of training facilities as a lower-cost alternative to ship-bridge simulators with visual scenes. Some blind-pilot simulators are configured so that they can be upgraded to include CGI.

In general, blind-pilot simulators are driven by the same mathematical model used for a ship-bridge simulator with visual CGI displays. Some radar simulations are also configured to appear as a small ship's bridge. These latter simulations are typically driven by less-sophisticated mathematical models than those used for full-mission or multi-task simulators.

Blind-pilot simulators that use the same mathematical model as a full-mission or multi-task ship-bridge simulator can be linked with them to enable multiple vessel interactions within the context of the training scenario. Because of the absence of the visual scene, the range of training that can be conducted is more limited than for a full-mission or multi-task ship-bridge simulator. Thus, although bridge team management training could be conducted on blind-pilot simulators, the full range of cues that prompt decision making and action are not present for all operating conditions. The training that occurs is often less complete and representative of human performance than is possible in a ship-bridge simulator.

Blind-pilot simulators are suitable for radar observer training and applied training in electronic navigation. Their utility for situational awareness is limited to conditions of reduced visibility because of the absence of visual scenes.

Special-Task Simulators

The substantial rise in computational power and the rapid spread of multimedia compact disk drives in desktop computers has stimulated considerable interest in the potential of these special-task simulator systems for training beyond traditional computer-aided learning.

Special-task simulators place the participant outside a functional mockup of a ship bridge or bridge equipment. The training environment produced is highly simplified and typically requires artificial interactions between student and simulator, such as controlling all operations through keyboard commands and data entries. The participant must conceptualize the ship bridge or bridge equipment more than in a full-mission or multi-task simulator. The simplified environment of a special-task simulator can provide a highly focused learning experience for specific nautical knowledge needs and tasks. The trainee may not, however, react in the same manner as when aboard ship because the manner and form of the stimuli and interactions differ greatly from actual operating conditions.

FIGURE 2-4 Elements of a sample PC-based simulator program. (Photo courtesy of PC Maritime, Ltd.)

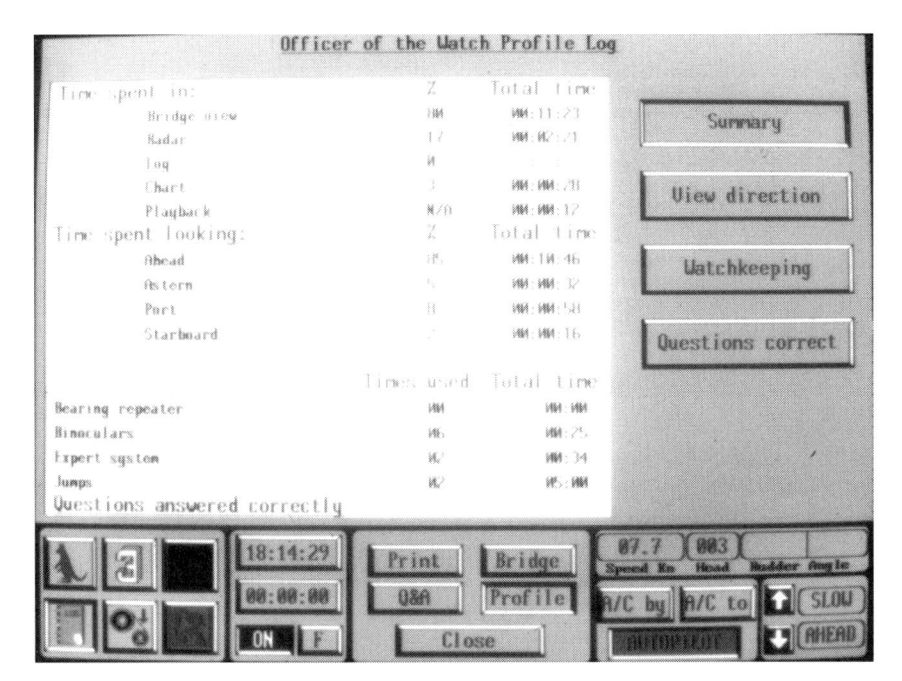

FIGURE 2-5 Elements of a sample PC-based simulator program. (Photo courtesy of PC Maritime, Ltd.)

Basic microcomputer desktop simulators normally consist of a single microcomputer, a single monitor or screen projection, and an input device, usually a keyboard. In contrast to the full range of conditions generated with full-mission ship-bridge simulators, the training environment created by a microcomputer desktop simulator is greatly simplified. Figures 2-4 and 2-5 illustrate elements of a sample PC-based simulator program.

The cue domain of the desktop training environment is not only greatly simplified from actual operating conditions, but also differs significantly in manner, form, and correctness of the presentation. For example, detail and accuracy of visual displays are generally incomplete. Visual display images are smaller than real life because details are compressed as a result of screen size, and only one sector (e.g., 90 degrees) can be viewed. There is also a single-data entry device (such as a keyboard, mouse, or specially configured device) and only visual representation of equipment controls rather than actual controls.

The operation of bridge equipment on microcomputers can be very artificial. Angular relationships are distorted for several reasons, including the relatively flat screen, the need to call up different screens through keyboard entries to view other sectors and instrument displays, and a lack of depth perception. One alternative, to display on several screens, can be costly.

Although angular relationships can be correctly displayed with a bird's-eye view, the display is even more restrictive than a visual display. Visual depth perception problems can be avoided when using radar emulations that have the ability to measure distances. Bird's-eye views that move a vessel "footprint" over an electronic chart or similar geographic display are very difficult to correlate with real operating conditions. This problem exists because of scaling factors, the mariner's lack of familiarity with the use of electronic charts (which are only now beginning to enter the commercial fleets), and the difficulties in correlating geographic representations that do not take the form of electronic charts with reference features actually used for navigation.

Manned Models

An early application of physical modeling of ships for training occurred in 1966 with the building of the world's first manned-model training facility in France. This facility was initially developed for training masters in the maneuvering capabilities and shiphandling procedures for very large crude carriers. At that time the size of these carriers represented a quantum jump in vessel size, with attendant changes in maneuverability. An engineering organization with extensive experience in port and harbor development was selected as the contractor and has operated the commercial facility since its opening (Graff, 1988). Three additional manned-model training facilities have been developed, one each in England, the U.S. Navy Amphibious Base in Little Creek, Virginia, and, more recently, Poland. The Little Creek facility, where a number of merchant mariners received shiphandling training in conjunction with Naval Reserve training, was closed by the U.S. Navy in 1993 as a cost-reduction measure.

Physical models, in contrast to ship-bridge and radar simulators, always simulate ship motions and shiphandling in fast time because of scaling factors. Manned models are believed by many to provide realistic representation of bank effects, anchor effects, shallow water, and ship-to-ship interactions. For manned models, the hull forms and water medium give a more realistic representation of important hydrodynamic forces acting on ships during typical maneuvers, at least for the particular ships modeled. Figures 2-6 and 2-7 are examples of manned-model simulators.

Manned models are well suited for shiphandling training over a wide range of operating conditions, limited principally by the resources and body of water available at the three existing commercial facilities. The models place the trainee inside a real operating environment, albeit at modified physical and time scales. Nevertheless, the behavior of the manned models, at their scale, creates training conditions that closely approximate operating conditions and vessel behavior at full scale.

Manned models are used primarily by experienced mariners who have the operational experience needed to adjust to the training environment and are able to adapt lessons learned to real-world applications. Less experienced mariners

FIGURE 2-6 An example of a manned-model simulator. (The Southampton Institute, Warsash Maritime Centre, United Kingdom)

may require additional model time to develop a frame of reference for transferring training insights to full-scale conditions.

Limitations of scale models (acknowledged by operators of manned-model facilities) include exaggerated effect of wind, restrictions on the number of different ship types and channel configurations, stereoscopic effects of human perception at a reduced scale, and significantly compressed operating time scale. In particular, the vessel size and time scaling must be adapted to during training and subsequently interpolated into skills and abilities in actual operations. Despite these concerns, there is wide support, particularly among pilots, for use of manned-model simulators.

Virtual Reality Training Systems

The latest development in simulator-based training is the emergence of virtual environment technology. A recent report (NRC, 1994b) states that:

> Virtual environment systems differ from traditional simulator systems in that they rely much less on physical mockups for simulating objects within reach of the operator and are much more flexible and reconfigurable. Virtual environment systems differ from other previously developed computer-centered

FIGURE 2-7 An example of a manned-model simulator. (SOGREAH Port Revel Centre)

systems in the extent to which real-time interaction is facilitated, the perceived visual space is three-dimensional rather than two-dimensional, the human-machine interface is multimodal, and the operator is immersed in the computer-generated environment.

Unlike existing ship-bridge simulators, virtual environment systems use helmets with a video display and sound capabilities, and its sensor systems detect movements of a person's extremities. Such systems, although currently somewhat limited in capabilities, are progressing toward more complete simulations of visual environments and toward development of better simulations of sound and feel. If successful in achieving realistic training environments, some virtual environment technologies have the potential to reduce simulator costs. Other virtual environment technologies may increase costs, but they may also significantly extend simulator capabilities.

The NRC report also found that "despite the enthusiasm and the 'hype' surrounding the synthetic environment (SE) field, there is a substantial gap between the technology available and the technology needed to realize the potential of SE systems envisioned in the various application domains" (NRC, 1994b). At this time it is unclear which of the diverse features of virtual environments will enhance training effectiveness. A major research program funded by the U.S. Navy is currently exploring this issue. The goal of this program, Virtual Environment for Submarine Pilot Training, is to develop, demonstrate, and evaluate the training potential of a stand-alone virtual reality-based system for deck officer training and to integrate this system with existing submarine piloting and training simulators (Hays, 1995).

The potential of virtual reality must be balanced against the fact that the current manner and form of virtual reality simulations are substantially different from the ship-bridge operating environment and normal context in which mariners perform. There is no research to determine whether or to what degree a realistic virtual environment simulation might be possible for the commercial marine operating environment. It also remains unclear whether enhanced technological capabilities of virtual environments will improve current levels of simulator effectiveness generally, and, if they do, whether use of these capabilities will be cost-effective. Given the state of the U.S.-flag merchant marine, it would seem impractical to invest in virtual environment research at this time. The results of research from other sectors might be adapted in the future.

USE OF MARINE SIMULATORS FOR TRAINING

The existing training methodology in the marine industry has evolved based on old technologies, developed as ships have developed—slowly, over a long period of time, in a conservative industry. Initially, the method for using simulators in training was as an addition or complement to existing programs.

Simulation enables creation of dynamic, real-life situations in a controlled classroom environment where deck officers and pilots can:

- practice new techniques and skills;
- obtain insight from instructors and peers;
- transfer theory to real-world situations in a risk-free operating environment;
- deal with multiple problems concurrently rather than sequentially; and
- learn to prioritize multiple tasks under similar high stress, changing conditions to those in actual ship-board operations.

Simulators can also be used effectively to bring a new dynamic into the classroom by combining books and lectures with real-time simulator-based instruction to teach rather than just explain real operating skills.

Although simulation can be a relatively low-cost option for training, use of simulation must be based on its suitability to training objectives. An expensive full-mission simulator for early instruction of navigation skills, for example, may be inappropriate if a less-expensive, limited-task simulator, or even interactive microcomputer-based instruction, would meet training objectives. A well-designed program of instruction will use a less-expensive, limited-task training device or interactive microcomputer-based system, designed to focus on specific tasks, rather than a full-mission device that is better suited for systematically integrating all performance components.

The recent growth of all forms of marine simulators, and particularly ship-bridge simulators, is driven substantially by technology. Simulator facility operators have taken advantage of rapid advances in computer computational capabilities and graphic imagery to enhance the training marketing potential. Smaller, faster microcomputers with greater memory capacity have made it technically and financially practical to drive desktop simulators with complex hydrodynamic models at a reasonable cost.

The latest advances in computational capabilities and software have permitted the addition of highly detailed visual scenes linked to full-motion platforms, providing a high level of apparent realism to marine simulators. However, whether these features achieve a sufficiently faithful reproduction of real-world effects, or add significant value to simulator-based training, has not been determined through either quantitative or qualitative assessments.

Motivations for Simulator-Based Training

Simulation as a Teaching Tool

Simulator-based training permits hands-on training to be conducted in a realistic marine environment without interfering with the vessel's operation or exposing it to risk. Training can continue independent of adverse weather conditions, vessel operating schedules, and other training conditions (e.g., harbors and waterways). The following sections discuss simulation characteristics as a teaching tool.

Safety. Risks associated with training on operational equipment are a concern in any industry. Within the commercial air carrier industry, the widespread use of simulators in training has reduced training accidents.

Simulators allow students to repeat a risky operation several times if needed. Unlike training on operational equipment, where an instructor must be prepared to intervene at all times, risky maneuvers can be safely practiced on a simulator. Simulation enables the placement of full responsibility on the prospective officer-of-the-watch before that individual actually assumes the duties of a licensed deck officer.

This teaching situation is different from that aboard a school ship, where a licensed deck officer is ultimately responsible for the vessel's safety and must necessarily intercede and tutor as appropriate to the situation. Intervention, or even just the possibility of such an intervention, can cause very significant differences among candidates for third mates in the level of confidence and ability to lead watchkeeping teams.

In on-the-job training, concerns for safety of the vessel might cause an instructor to intervene earlier than is desirable for efficient progress of learning. During real operations, it may be necessary to interrupt training to avoid a real-life accident. In simulator-based training, the instructor can allow students to make mistakes, to see the consequences, and possibly to practice recovery procedures. Although there is limited objective evidence on the value of permitting students to make and recover from errors, many instructors believe students can receive full benefit from the training event only if given the freedom to make mistakes.

Lesson Repetition. Using simulation, the instructor can terminate a training scenario as soon as its point has been made or repeat it until the lesson has been well learned. In contrast, opportunities for repetition are very limited during actual at-sea operations; the opportunity to repeat an exercise in on-the-job training aboard ship may not occur for weeks or months.

Recording and Playback. Another feature of simulator-based training is the ability to record and play back the just-completed scenario for review, evaluation, and debriefing purposes. As a teaching tool, recording and playback empower the instructor to let mistakes and accidents happen for instructional emphasis and allow trainees to review their actions and their correct and incorrect decisions and experience the results of their performance after the exercise is completed. As an assessment tool, recording and playback can provide a history of performance that serve as a "second opinion" if a candidate challenges an assessor's opinion, thereby minimizing what might be an otherwise subjective licensing assessment.

Flexibility. Simulator-based training permits systematic scheduling of instructional conditions as desired by the instructional staff or as directed in the training syllabus. Simulation permits the use of innovative instructional strategies that may speed learning, enhance retention, or build resistance to the normally disruptive effects of stress.

Multiple Tasks and Prioritization. Deck officers at all levels of responsibility must continually decide at any given time, in any given situation, which among a number of tasks are most important. Before simulator-based training, a new officer's initial training often consisted of a range of skills that were taught, practiced, and examined separately. Use of simulation in training programs makes it possible to transfer classroom skills and to practice and prioritize multiple tasks simultaneously. Simulation training enhances development of skills and provides the opportunity to exercise judgment in prioritizing tasks.

Training on New Technologies. By employing features such as the ability to repeat training exercises and to record and play back performance, simulators can provide a safe environment for training mariners in the use of new equipment. For some new equipment it is possible to place desktop simulators on board ships to provide an opportunity for independent training.

Peer Interactions. Simulator-based training at simulator facilities can provide a forum for peer interactions and evaluations that might not otherwise occur. Because of the often solitary nature of their work, masters and pilots can routinely serve for years without having their work observed or critiqued by their peers. Simulator-based training can provide an opportunity for these mariners to improve their competency and learn new techniques by having old habits challenged and corrected in a safe environment.

Cost Effectiveness

Although the most obvious goal of using simulation is improving performance, cost effectiveness is also important. Simulators in the commercial air carrier (see below) and marine industries generally (although not invariably) cost less to build and operate than the operational equipment being simulated. The commercial air carrier industry is able to conduct transition training to a new aircraft entirely in simulators and at substantial savings over costs of the same training conducted entirely in an actual aircraft.

In the marine industry, any calculation of savings comparing cost of using commercial ships solely as training platforms with those of simulators are almost entirely speculative. Training aboard commercial ships can be difficult or, in some cases, impractical because of risk, operating practices, and schedules. In addition, continuing training is mandatory in the commercial air carrier industry, while it is not in the marine industry. Without mandatory requirements, some shipping companies will not finance simulator-based training.

The Need to Reinforce Simulator-Based Training

The committee found that once a simulator-based training course has been completed, learned knowledge, skills, and abilities are generally not systematically reinforced. However, the committee also found examples in which simulator-based training programs were integrated with actual operations or followed up with onboard training. In cadet training, there is an ongoing research effort to reinforce simulator training. The Maritime Academy Simulator Committee[3] has designed a shipboard-experience profile survey form. The form, which is to be completed by each cadet after completion of his or her training cruise, commercial sea year, or simulator course, is an attempt to reinforce simulator-based training.

Some shipping companies reinforce simulator-based training. For example, in one shipping company the instructor who conducted simulator-based training for company employees subsequently conducted onboard training "simulations"

---

[3]The Maritime Academy Simulator Committee is composed of two representatives from each maritime academy and one representative each from the U.S. Maritime Administration and the U.S. Coast Guard. The purposes of the committee are to (1) determine common standards for certification of simulators, programs, and instructors; (2) develop a standardized curriculum for ship-bridge simulator-based training of cadets; and (3) determine sea-time equivalency ratios for full-mission ship-bridge and limited-task simulators.

(i.e., drills) to extend theory into practice. Another shipping company has a program that relieves an entire bridge team from duty, flies them to a simulator facility for specialized training, and returns them to their vessel where the lessons can be immediately applied. This approach provided a climate for constructive change because the vessel's master and all deck officers were exposed to the same concepts. Company management involvement in the training program demonstrated its commitment to the application of training results.

The training of an entire bridge team (including engineering officers) as a unit can mitigate the lack of universal standardized procedures for bridge team operations (as opposed to standing orders aboard individual vessels), a common situation that constrains transfer of training lessons to actual operations. Standardization can occur because the bridge team trains together, is provided the same training, and is kept together as a team.

Training Limitations of Desktop Computers

Desktop computers normally provide a single workstation, designed for use by one person at a time, even if linked into a training network. Trainees are isolated and do not interact with other bridge team members or pilots. Trainees can also be isolated from the instructor, particularly if training is not conducted at a training facility. The desktop training environment provides neither the same cues nor instructional oversight common with a ship-bridge simulator.

Instructional oversight can be improved for desktop training by placing microcomputer simulation workstations in a classroom or laboratory and linking them to an instructor control station with diagnostic capabilities. This approach has been adopted by several manufacturers of software for rules-of-the-road desktop training simulators. Providing the diagnostics at each workstation may enhance individual training; it is not, however, a substitute for student-instructor interaction and debriefings.

Microcomputer desktop simulators present a very artificial training environment compared to the latest generation of ship-bridge simulators. The manner of stimulating human performance is substantially different. These differences do not mean that desktop simulators lack training value, but that the capabilities and limitations of desktop simulators need to be understood. The committee could not find any data or research that investigated whether these differences significantly affect training outcomes. In particular, an understanding of transfer effectiveness is less well developed for microcomputer-based simulators than for ship-bridge simulators.

The simplified training environment of desktop simulators may be modified by physically separating input devices to require participant movement, involving several individuals in the simulation, or by including multiple monitors. Several training system companies have developed software and entry devices that make it possible to emulate specific nautical systems, such as radars and global maritime distress safety systems. Capabilities of some of these systems approach small-scale, limited-task simulators.

Examples of Training not Currently Appropriate for Ship-Bridge Simulators

Heavy-Weather Training

With current technology, some tasks for operating in heavy weather (e.g., shiphandling) are learned more effectively at sea than in current-generation computer-based or manned-model simulators that are not capable of re-creating the motion experienced by ships in heavy weather. Some other heavy-weather tasks (e.g., steering) are more effectively learned aboard a rolling and pitching ship, where actual forces of the ship working in the sea can be felt. Currently available motion-base simulators are not adequately validated for complex interactions among steering, heavy seas, and wind.

Shiphandling Training for Pilots

There is general agreement among pilots that simulators are very useful for teaching shiphandling principles and theory for enhancing pilot skills, for providing a forum for pilots to teach new or refined techniques to each other, and for developing and testing methods for handling new ship types or operating under changing port conditions (NRC, 1994a). There is also, however, a consensus among pilot organizations that full-mission ship-bridge simulators are not as effective for instructing pilots in underway close-quarters shiphandling.

Pilots routinely learn shiphandling skills during the course of their work from senior pilots in one-on-one, hands-on training experiences. Pilots effectively use the ship for training. Changes in the industry have not significantly altered this situation. With virtually no exceptions, simulator-based or other training is not substituted for required trips over a pilotage route for familiarization or shiphandling training for independent marine pilots, although the USCG does, in special situations, allow such a substitution for federal pilot endorsements.

SIMULATION IN MARINE LICENSING

Refinements to improve the marine certification process with respect to STCW guidelines are under way, as are several USCG-initiated measures to improve U.S. marine licensing. In particular, comprehensive guidelines for application of simulation in marine certification are under development by the IMO, an initiative supported by the USCG. These changes are being approached within the domain of existing sea-time requirements. The practice of unstructured learning during sea time as a basis for professional development has not been quantitatively examined by the IMO or the USCG to determine whether this method satisfies modern marine safety needs or results in properly trained mariners. Nevertheless, there is growing recognition that existing practices are not optimum for modern operating conditions (see discussion in Chapter 6).

In 1993 the USCG appointed a "focus group" consisting of USCG officers and civilian employees familiar with marine licensing and commercial or USCG operations to study the licensing program and make recommendations for changes. The group's report, *Licensing 2000 and Beyond* (Anderson et al., 1993), contains a number of recommendations concerning increased use of simulators for training and licensing mariners. The report includes broad-reaching, often innovative, and controversial concepts. It is, however, limited with respect to the inclusion of facts and analysis.

The focus group observed that "although sea service experience and static testing techniques have formed the traditional basis for the seaman's training, licensing, and the certification process, this arrangement alone no longer provides the best methodology for ensuring professional proficiency." They concluded that "given the pace and sophistication of technological change within the industry and complexity of the affected trades, a mixture of comprehensive training and sea service experience offers the best opportunity to achieve higher standards of the professionalism and casualty reduction goals."

The report recommends that the USCG adopt fundamental, higher-order principles to guide the agency's marine licensing program, including a stronger, more focused role in establishing competency standards and determining competency. The report also recognizes that use of simulators to test more than definitive subject knowledge—for example, as a "road test" of individual abilities—has not been fully demonstrated.

The focus group explicitly noted that there is a mismatch between requirements for radar observer certification and actual use of radar. The group noted a need for a well-developed course-approval methodology, including adequate training for USCG personnel involved in overseeing review, approval, and maintenance of approved courses.

With respect to sea-time requirements, the report opined that "proper utilization of the approved course concept would allow reductions in actual sea service experience, enhancement of professional proficiency and in many instances elimination of the U.S. Coast Guard's examination process." The report also strongly stated that "until such time as the quality and integrity issues surrounding 'approved courses' are resolved, no further move towards reliance on such courses should be contemplated."

The focus group recognized that its findings reflected a USCG perspective and recommended submitting the report to the USCG's Merchant Personnel Advisory Committee (MERPAC) for validation. The report was referred to MERPAC at its December 1994 meeting.

Current Use of Simulation in USCG Licensing

The USCG has been involved in the development and approval of training courses for merchant mariners for over half a century. The agency's role has evolved to one in which it approves courses used to satisfy international and national marine certification and licensing requirements, including courses where some limited remission of sea service is granted. The USCG has, in the past few years, expanded its concept of using training and associated course approvals to include training courses that could be voluntarily substituted for an examination. Required radar endorsements are an example. In one specialized case, the agency allows substitution of specialized training for some round trips required for a federal first-class pilot's license or endorsement on selected port entry routes in the Great Lakes. By linking training to examination requirements, the USCG has broadened its domain of course approvals.

Radar Observer Certification

The STCW guidelines require candidates for certification as masters, chief mates, and officers in charge of navigation for oceangoing ships over 200 gross registered tons to demonstrate knowledge of the radar fundamentals, operation, and use for navigation and collision avoidance purposes (IMO, 1993). For this demonstration the candidate uses a radar simulator or, if not available, a maneuvering board.

The USCG has promulgated regulations requiring radar observer certification to meet STCW guidelines. This certification uses radar training courses involving examinations of knowledge and practical simulator demonstrations. Courses must be approved by the USCG to meet the regulatory requirement. To receive a certificate, candidates must successfully complete a curriculum including fundamentals, operation, and use of radar; interpretation and analysis of radar information; and plotting.

Course Approvals and Sea-Service Credit

Subpart C of 46 CFR Part 10 contains general USCG guidelines to approve instruction courses. Currently, some approved courses are accepted as partial credit toward required sea service. Among these are a small number of courses that use a simulator as one of the instructional tools. For time spent in approved, full-mission simulator-based courses, the USCG presently grants equivalent sea-service credit for a portion of the license upgrade required sea service (USCG, 1993).

Courses are reviewed by the USCG prior to approval and are periodically audited thereafter, subject to the availability of resources. Each course is individually evaluated for sea-time credit to be granted. "Simulator training cannot be substituted for recency requirements, but can be substituted for a maximum of 25 percent of the required service for any license transaction" (46 CFR 10.305).

The USCG grants some equivalency for simulator-based training in lieu of service leading to endorsements for special operations, such as service as offshore oil port mooring master and competence certificates for liquid natural gas carrier officers. Some of the specific criteria for USCG course approvals are described in Appendix C.

Recent Changes in USCG Use of Simulation in Licensing

In November 1994, the USCG accepted an unsolicited proposal from the SIMSHIP Corporation to conduct a training course that would combine training, written examination, and ship-bridge simulator assessment. This course will be equivalent to the written examination for the master's unlimited oceans license and may be selected by the applicant as an alternative to the written examination (see Appendix F).

The use of combined training and examination is a hybrid. Licensed master mariners employed by the simulator operating company (rather than USCG license examiners) are authorized to perform simulator-based training and conduct portions of the examination using simulations.

Approval of this course represents one of only two instances (the other being the radar observer certification) in which the USCG has delegated its role as licensing examiner. The efficacy of this approach and adequacy of agency resources to oversee and implement these programs are issues of concern within the marine community because, as currently structured, none of the master's license program elements—training instructors, licensing assessors, simulator platform, or training and testing scenarios—have been validated.

Approval of this course does, however, represent an important opportunity for the USCG to conduct significant research in the use of simulators for evaluation and assessment. As a part of its program oversight, the agency should require the simulator facility operator to document all changes made to the primary program elements, including the training course curriculum and the simulator scenarios. The agency should collect and analyze this and other data concerning factors such as transfer effectiveness and long-term performance of candidates who successfully complete the course. The data collected could be used to revise and improve the USCG licensing program.

USE OF SIMULATION IN NONFEDERAL MARINE LICENSING

State pilotage authorities are responsible for the establishment of training and license requirements and implementation criteria or standards. None of the state authorities currently require pilots to participate in marine simulator-based training or evaluations as a condition of original licensing. Several pilot associations have begun to incorporate marine simulation into their apprentice programs as a supplement to existing training requirements. A growing number of state pilotage authorities and pilot associations have established continuing professional development requirements that require specific training, such as manned-model shiphandling courses, simulator-based courses, or bridge resource management using interactive videos or other training media.

COST OF SIMULATOR-BASED TRAINING

The cost of simulator training varies widely, depending on the simulator facility used, the requirements of the particular training program, and the travel, housing, and food expenses for trainees. The committee was able get actual cost figures from several facility operators and developed the following general cost information:

- Most computer-based simulator training programs appear to cost $500 to $700 per day per student.
- Manned-model simulator shiphandling training courses cost approximately twice that amount, or $1,000 to $1,400 per day per student.
- Many simulator facilities provide housing and meal facilities, which are available at a moderate cost to students. Others have cooperative arrangements with nearby hotels and motels. Lodging and meal allowances are at least $100 to $150 per day per student.
- Travel costs are the largest variable, since the training facility can be located across the country or outside of the United States.

The committee visited East Coast facilities where West Coast-based students were undergoing training. In another case, a Northeast-based facility was training students who had traveled from the Gulf Coast. Several pilot associations sent candidates for shiphandling training to European manned-model facilities. In contrast, cadets undergoing training at various maritime academies are all in residence.

The cost of the training is also affected by the specific training requirements of the company. Some commercial fleet operating companies require nongeneric, port- and ship-type-specific simulation models with customized simulator scenarios. These one-time development costs can be significant and are in addition to the general estimates of the costs noted above. Companies and individual students that can use a facility's training programs, with generic ship types, ports, and relatively standard scenarios are not subject to these additional development costs.

These costs of training are clearly significant. For the most part, the costs are paid by the employer, either directly or through employer contributions to union training funds. The committee also found that some facilities attempt to offer special rates to individual students who do not have an employer affiliation. However, for most mariners, the cost of simulator training is considered beyond the reach of those who lack a union or employer sponsor.

SIMULATION IN THE COMMERCIAL AIR CARRIER INDUSTRY

In using simulators for training and certification, the application of simulators in the commercial air carrier industry represents a state of the art. The modern aircraft simulator is an invaluable resource for commercial pilot training and certification, due in part to the influence and direction of the Federal Aviation Administration (FAA).

The operation of commercial aircraft differs significantly from commercial ships with respect to operating environment, operating platforms, and professional regulation. The regulatory concepts used in the commercial air carrier industry differ substantially from marine transportation. Professional certification in the commercial air carrier industry is platform-specific, whereas marine certification is necessarily much more generalized. In addition, as discussed in Chapter 1, the duties and responsibilities of a marine deck officer are very broad, ranging from watchkeeping to conducting ship's business. Commercial air carrier pilots, in general, have a much narrower range of responsibilities. Despite these differences, it is possible to identify concepts and frameworks within the commercial air carrier system that could be adapted and applied to the marine industry.

Airplane Simulators for Certification and Training

As simulators have increased in ability to behave like aircraft, the commercial air carrier industry and the FAA have increased the permissible amount of training and checking accomplished in a simulator. Use of flight simulators in certification of commercial air carrier pilots is voluntary, with the single exception of low-level windshear training, a particularly critical emergency maneuver which is part of recurrent training requirements. All other flight maneuvers required as part of recertification may be carried out in a flight simulator or in an actual aircraft. The philosophy behind this substitution for specific levels of training and checking is an assessment of simulator fidelity. In general, more critical maneuvers must be conducted either in a real aircraft or in a simulator that looks, feels, and behaves like one.

The earlier (late 1960s), relatively inflexible, regulatory standards for airmen focused on pilot technical performance in individual duty positions (e.g., captain, first officer, flight engineer). Regulations concerning pilot qualifications were explicitly oriented around the individual's capacity to perform the duties of the position without assistance. Training and checking focused on the captain. Considerably less attention was given to training and checking other crew members, with no particular requirement to provide training of individuals to operate as a crew.

In 1988 the U.S. National Transportation Safety Board (NTSB) issued Safety Recommendation A-88-71, directing a review of all initial and recurrent flight crew training programs to ensure that they included simulator or cockpit exercises involving *cockpit resource management*.[4] The original motivation for the recommendation was the result of a study that found that at least 60 percent of commercial air carrier accidents could be attributed to some form of preventable crew error (Longridge, n.d.).

In 1994 the NTSB strengthened its call for crew resource management by proposing that all U.S. Part 121[5] carriers be required to provide their flight crews with a standardized, comprehensive crew resource management program, focusing on decision making and challenging another crew member's errors. The impetus behind this second action was a study of 37 major U.S. air carrier accidents between 1978 and 1990. The study found that, of the 302 specific errors identified in these accidents, the most common were related to procedures, tactical decisions, and failure to monitor or challenge another crew member's error (NTSB, 1994). These causes closely reflect the causes identified by the NTSB in marine industry accidents cited in Chapter 1.

The Advanced Qualification Program

In 1991, partially in response to the 1988 NTSB recommendation, the FAA established the Advanced Qualification Program (AQP). The program provides certified air carriers and training centers with a voluntary alternative for training, qualifying, and certifying regional and major air carrier crew performance. The AQP offers U.S. air carriers the flexibility to tailor training and certification activities to a carrier's particular needs and operational circumstances.

The AQP is distinguished by its emphasis on proficiency-based training and qualification. Each AQP applicant, rather than the FAA, develops its own set of proficiency objectives, based on a level of job-task analysis sufficiently detailed to justify substitution for traditional training and checking requirements. The analysis required to apply for AQP is substantial. The focus is on identifying the frequency at which essential knowledge, skills, and abilities are reinforced during actual operations and the rates at which they degrade. Once these factors are

---

[4]Broadly interpreted, cockpit resource management means the effective and coordinated use of all resources available to crew members, including each other.

[5]Aircraft flying under Part 121 of the Code of Federal Regulations 14 (Aeronautics and Space) are domestic, flag, and supplemental air carriers and commercial operators of large aircraft (more than 30 seats, 7,500 pounds payload).

determined, it is anticipated that AQP applicants will be able to conduct interim refresher training using less costly alternatives in place of the highest level of visual flight simulation.

Refresher training is directed toward knowledge, skills, and abilities identified in the analysis as degrading more quickly. Such skills must be restored more quickly than now occurs through annual recertification. This approach would in turn enable the AQP applicant to reduce recertification frequency through a visual flight simulation course.

The arduous job-task analysis required by the FAA to qualify for AQP is expensive—one airline spent over $1 million to develop its analysis. The prospective reduction in recertification training costs, however, was anticipated to recover the analysis cost in less than two years.

Instructor Qualification

Flight simulator instructors are required to be licensed pilots to ensure their technical competence and credibility with the airmen being trained. Each instructor and checker is required to participate actively in either an approved, regularly scheduled line-flying program as a flight crew member or in an approved line-observation program in the airplane type for which that person is instructing or checking. Each instructor and checker is given a minimum of four hours training to establish familiarity with the simulator facility operator's advanced simulator-based training program. Instructions include training policies, instruction methods, simulator controls operation, simulator limitations, and minimum equipment required for each training course.

Air Carrier Simulator-Model Validation

Validation of an aircraft simulator model typically involves the following four levels (though the actual process may vary by facility) (NRC, 1992):

1. individual modules or subprograms,
2. small packages or groups of subprograms related to functionality (e.g., propulsion systems),
3. dynamic response verification of the complete mathematical model, and
4. pilot-in-the-loop, which allows for some minor adjustments to models to improve realism of simulated pilot cues.

Air Carrier Simulator Classification and Qualification

The evolution of simulator technology and the increase in use have required a similar evolution in the thoroughness and complexity of simulator qualification criteria. Simulators in the commercial air carrier industry are divided into four designated levels. Complexity of the highest level is not required of all simulators. There are also a series of "training-device"[6] levels, which are generally nonvisual or limited-task training tools. The FAA has published criteria for:

- qualifying a simulator or training device and determining the qualification level,
- determining what training and checking is allowed for each level simulator, and
- calculating credits allowed for training-device use in the completion of specific training and checking events in air carrier flight training programs and in pilot certification activities.

Simulators are evaluated, qualified at a particular level, and recurrent evaluations are required every four months through the National Simulator Evaluation Program (NSEP). Simulators are evaluated as objectively as possible. Performance and handling qualities are evaluated according to the engineering specifications of the actual aircraft being modeled (NRC, 1992), and pilot acceptance is determined through a subjective validation test conducted by a qualified FAA pilot. The NSEP specifies exact maneuvers and tests to be performed and acceptable tolerance limits for each test.

Differences Between Air Carrier and Marine Simulators

The use of simulators in the commercial air carrier industry reflects, in part, different operating environments, practices, and "fleet" composition. In the certification of simulators, there are several fundamental differences between visual flight and ship-bridge simulators. For example, visual flight simulators for commercial air carriers are linked directly to development of specific airframes and are not modified to permit training in multiple airframes (NRC, 1992). This practice is possible because of the large numbers of like airframes owned and operated by commercial airlines.

In contrast, ship-bridge simulators are not only developed independently from the vessels they simulate, but are also routinely used to train in multiple hull forms and sizes. As a result, some marine simulator facilities either use a number of models to meet the specific application needs of training sponsors or adjust their models to simulate a number of different vessel types or sizes.

---

[6]An airplane flight training device is a full-scale replica of an airplane's instruments, equipment, panels, and controls in an open flight deck or an enclosed airplane cockpit, including the assemblage of equipment and computer software programs necessary to represent the airplane in ground and flight conditions to the extent of the systems installed in the device; it does not require a force (motion) cuing or visual system.

Unlike commercial air carrier simulators, there are no industrywide standards for marine simulators. Marine simulators vary greatly in mathematical hydrodynamic models, scenario databases, and algorithms. The practice by some simulator operators of adjusting simulator models can cause problems. Problems with marine simulator models—and research needed for measuring and assuring the validity of ship performance models and scenarios—are discussed in Chapter 7 and Appendix D.

In training on simulators, the concept of bridge team and bridge resource management in the marine industry is similar to that of cockpit resource management. Also, the concept of using all levels of simulators, special task through full mission (as practiced in commercial air carrier training) for the progressive development of knowledge, skills, and abilities could be adapted to marine operations. The combined training and assessment approach approved by the FAA would, however, be very difficult—if not impractical—to implement in the marine sector.

REFERENCES

Anderson, D.B., T.L. Rice, R.G. Ross, J.D. Pendergraft, C.D. Kakuska, D.F. Meyers, S.J. Szczepaniak, and P.A. Stutman. 1993. Licensing 2000 and Beyond. Washington, D.C.: Office of Marine Safety, Security, and Environmental Protection, U.S. Coast Guard.

Drown, D.F., and I.J. Lowry. 1993. A categorization and evaluation system for computer-based ship operation training. Pp. 103–113 in MARSIM '93. International Conference on Maritime Simulation and Ship Maneuverability, St. Johns, Newfoundland, Canada, September 26–October 2.

Graff, J. 1988. Training of maritime pilots—the Port Revel viewpoint. Pp. 62–76 in Proceedings of Pilot Training, Southampton, England, July 12–13.

Hays, R. 1995. Personal communication to Committee on Ship-Bridge Simulation Training, National Research Council, describing Navy's virtual environment for submarine piloting training.

IMO (International Maritime Organization). 1993. STCW 1978: International Convention on Standards of Training, Certification, and Watchkeeping, 1978. London, England: IMO.

IMO News. 1994. World maritime day 1994: better standards, training, and certification—IMO's response to human error. IMO News (3):i–xii.

Longridge, T.M. n.d. The Advanced Qualification Program: Matching Technology to Training Requirements. Washington, D.C.: Federal Aviation Administration.

MacElrevey, D.H. 1995. Shiphandling for the Mariner, 3rd ed. Centreville, Maryland: Cornell Press.

Muirhead, P. 1994. World Maritime University, personal communication to Wayne Young, Marine Board, September 20.

NRC (National Research Council). 1992. Shiphandling Simulation: Application to Waterway Design. W. Webster, ed. Committee on Shiphandling Simulation, Marine Board. Washington, D.C.: National Academy Press.

NRC (National Research Council). 1994a. Minding the Helm: Marine Navigation and Piloting. Committee on Advances in Navigation and Piloting, Marine Board. Washington, D.C.: National Academy Press.

NRC (National Research Council). 1994b. Virtual Reality: Scientific and Technological Challenges. N.I. Durlach and A.S. Mavor, eds. Committee on Virtual Reality Research and Development. Commission on Behavioral and Social Sciences and Education. Washington, D.C.: National Academy Press.

NTSB (National Transportation Safety Board). 1994. A Review of Flightcrew Involved, Major Accidents of U.S. Air Carriers, 1978 through 1990. Safety Study NTSB/SS-94-01. Washington, D.C.: NTSB.

USCG (U.S. Coast Guard). 1993. Written communication from USCG OPA 90 staff to Committee on Ship-Bridge Simulation Training, National Research Council, July 23.

3

Effective Training with Simulation: The Instructional Design Process

Many current marine education and training approaches are outgrowths of an established profession with strong, traditional professional development practices. These practices reflect an approach to instruction in which instructors often served as both teacher and mentor, and ship's officers were trained to be generalists.

The use of ship-bridge simulators is becoming an accepted method of training in the international marine industry. Yet even as more simulators are being used, their use has not dovetailed smoothly into comprehensive training programs. Many simulator-based training courses were developed ad hoc, often designed to individual requirements of a shipping company or training establishment.

Among the experts the committee consulted, there was strong evidence that they had thought seriously about training needs and had organized their programs to address those needs. There were also diverse opinions on the exact character of those training needs. In some instances it appeared that training emphasis was guided by equipment capabilities such as simulators. The committee concluded that professional training in the maritime industry could be improved by effective advancement and systematic application of instructional design concepts.

DEVELOPING AN EFFECTIVE TRAINING PROGRAM

A simulator does not train; it is the way the simulator is used that yields the benefit. "It is easy to be impressed by the latest, largest full-mission simulator, but what is more important than the technology is how educational methodology is applied and whether it increases training effectiveness significantly, incrementally, or at all" (Drown and Mercer, 1995).

Trainees taking part in most simulator-based training courses can be divided loosely into two groups:

- unlicensed cadets who work through a series of structured courses; and
- fully qualified, licensed mariners who take stand-alone courses for updating, refreshing, and refining skills.

Although instruction design theories can be applied to both groups, the training programs and corresponding instructional processes will vary by trainee level.

An effective training program addresses the student's training needs with respect to knowledge, skills, and abilities. It exploits all media, from personal computer-based training to limited-task and full-mission simulators and applies the appropriate training tool to the specific level of training. For example, it would not be necessary to use a full-mission simulator for early instruction in rules-of-the-road training. Rather, a systematic approach to training promotes convergence toward full-mission expertise by developing basic modules of skills in several steps. This approach encourages the assembly of ever-larger skills modules until the trainee can exploit training on a full-mission simulator.

The instructional process is central to the overall focus of this report. Instructional design is a relatively new process. It has not advanced sufficiently to the point where this committee can provide a complete vision of how it should be implemented. The committee can, however, provide guidance.

Instructional design is an iterative process whereby training managers continually test innovations and improve training. It is an incremental approach that involves inserting new pieces developed by the instructional design process into existing training programs, assessing results, and then revising the program as necessary. This systematic application yields simulator-based training programs with clearly defined objectives, carefully designed training and evaluation scenarios, and qualified instructors. Figure 3-1 illustrates the iterative nature of this training process.

There are several stages to implementing the instructional design process (see Box 3-1). A vital first stage is *determining training needs*. This stage is important because current national and international licensing and certification requirements and guidelines focus primarily on knowledge rather than skills and abilities needed to effectively apply knowledge (Froese, 1988). Training needs can be developed by identifying gaps or missing elements between the trainee's required and actual knowledge, skills, and abilities.

The second stage is to *determine specific training objectives* (i.e., goals). Objectives identify each attitude, skill, and block of knowledge the trainee should have on successfully completing the course (Drown and Mercer, 1995). Once the course has started, these objectives should be clearly stated to orient the trainee. Development of training objectives should also include *developing performance*

EFFECTIVE TRAINING WITH SIMULATION

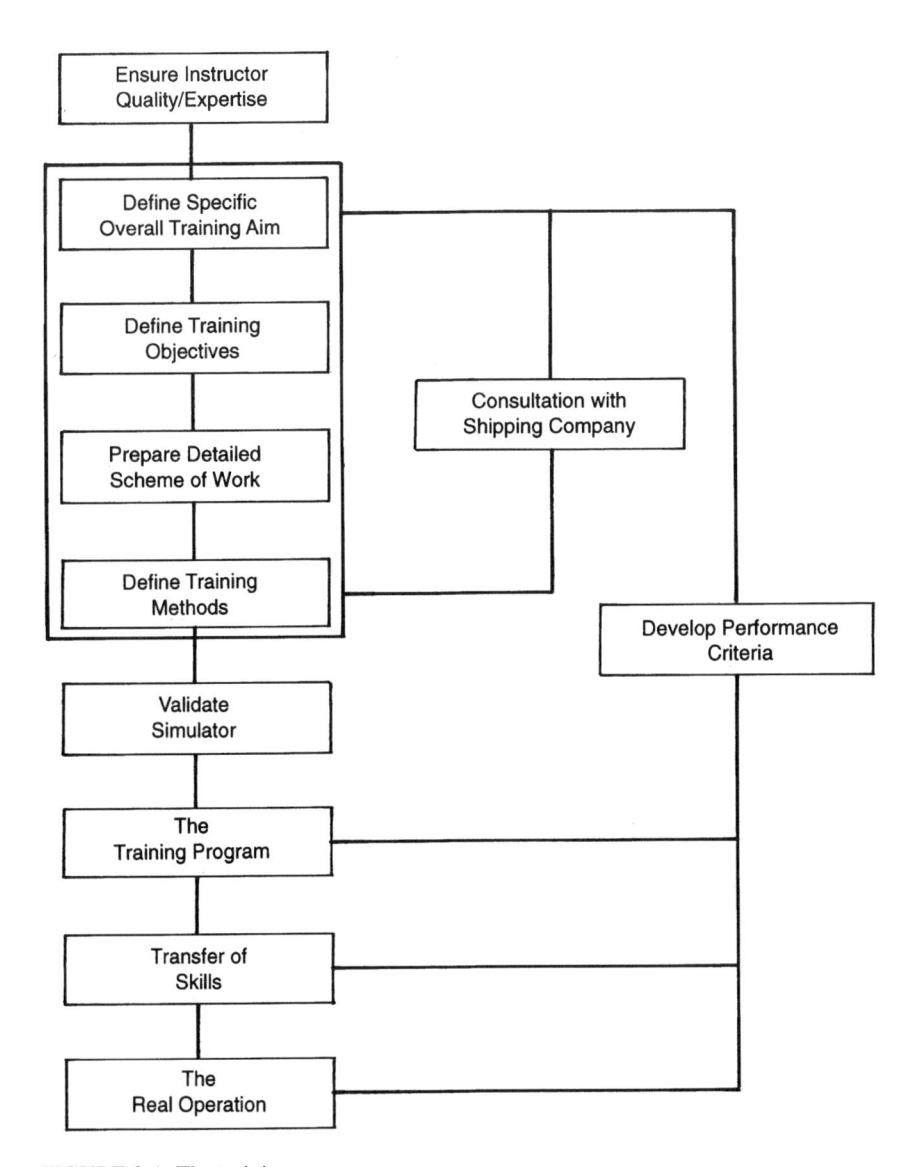

FIGURE 3-1 The training process.

> **BOX 3-1**
> **Elements of Instructional Design Process**
>
> - characterizing training populations and identifying specific training needs;
> - determining training objectives (i.e., goals);
> - determining course content and course material requirements needed to meet training objectives;
> - determining training methods;
> - identifying training resource requirements and correlating them with training objectives;
> - matching specific instructional techniques to curriculum content;
> - identifying student assessment requirements and developing assessment methodologies; and
> - establishing instructor qualification, selection, training, and certification requirements to ensure quality of instruction and successful curriculum implementation.

measures for determining whether or to what degree trainees have achieved the training objectives. These important elements of instructional design have not been well addressed and have sometimes even been neglected in simulator-based courses.

The third stage is to *determine the training methods* to be used. This stage includes an assessment of whether simulation use is relevant to achieving the training objectives. Assuming simulation is to be used, two things must be determined:

- the level of simulation (i.e., level of realism with respect to simulator components [see Chapter 4]) required to achieve training objectives (Hays and Singer, 1989) and
- the type of simulator—full mission, multi-task, limited task, or special task (see Box 2-1)—that will be of most value.

The development of the training approach should also consider factors such as:

- the total training program of which the simulator-based training is a part (e.g., cadet training toward a first license),
- trainee experience,
- type of training media,
- instructor's qualifications and experience, and
- cost benefit and effectiveness of the training program.

Once it has been determined that a simulator-based course is relevant to training needs, it is necessary to *develop a detailed course outline*. Finally, there needs to be a *validation of the simulator, the simulation, and the curriculum* to ensure relevance and suitability.

Anecdotal evidence, as well as experience and observations of the committee, suggests that differences in instructional techniques can result in a significant range of material that can be covered. The way material is covered also affects the relative value of the learning experience. These factors may be affected by simulator features and fidelity; however, limitations in these areas can be minimized or offset to a large extent for certain instructional objectives. For example, committee members found that, for bridge team and bridge resource management training, creative instructional design can be used to compensate for limitations in simulator capabilities.

According to a group of mariner instructors, many of whom met with the committee, the practical considerations shown in Box 3-2 are particularly relevant when structuring a mariner training program. These observations generally correspond with the results of human performance research.

APPLYING INSTRUCTIONAL DESIGN

Defining Training Needs and Objectives

The Trainee Population

There is great diversity in the professional backgrounds and maturity of trainees. Group members may range from entry-level to mates and masters to marine pilots and shore-based management officials. This diversity can affect the development of effective training programs because of the possible range of training needs—from entry-level training; license upgrades and renewals; refresher training; and familiarization training for specific ports, routes, vessels, or vessel types. For these reasons, simulator-based courses will be most useful when developed systematically.

Job-Task Analysis

The systematic definition of training needs requires a detailed understanding of tasks and subtasks necessary to perform the function to be trained. Traditionally, professional development aboard commercial ships and tugs has relied predominantly on "modeling the expert" for complex cognitive tasks—the person undergoing training watches and imitates the performance of senior professionals. This modeling is generally accomplished through on-the-job observation and hands-on experience aboard vessels. In the absence of comprehensive instructional design and attention to instructional abilities of the expert who is being modeled, this approach may have important limitations with respect to the quality of the learning experience.

> **BOX 3-2**
> **Training Insights from Mariner Instructors**
>
> - The overall training program, not just one component, must be effective.
> - Simulators should be selected and simulations designed to meet training needs rather than structuring training to fit the simulator.
> - Training performance should be measured against predefined performance criteria.
> - Training (or remedial training) should be continued until the required proficiency level is reached.
> - No matter how well the training program is designed, refresher training may be needed to maintain a level of knowledge and skills.
> - Careful, structured evaluation of trainee performance (a) prior to, (b) at the conclusion of, and (c) after the program (in the workplace) is necessary to monitor program effectiveness.
> - Conditions and attitudes in the workplace must be conducive to transfer of training. Policies, practices, and attitudes of regulatory authorities, shipping companies, and ships' masters are vital to ensuring effective training and subsequent transfer to the workplace.
> - The trainee's real-world performance, not the speed of acquisition of a task during training or the level of performance at the end of training, should be the measure of program effectiveness.

There are two obvious difficulties in using direct modeling for complex cognitive tasks. First, the rationale for the performance of the tasks is not only opaque to observers, but may also be implicit for the experts: they may not be able to describe their own thought processes or the rationale for them, even though they can perform the tasks. Second, in order to properly coach a novice, an expert may have to formulate an accurate mental model of the novice's understanding of the task (sometimes called the student model). But a novice's understanding of a task is not always obvious to the expert. These two key problems . . . suggest that there might be a limitation on the extent to which direct modeling of complex cognitive skills can be done (NRC, 1991).

The instructional design process offers an alternative to on-the-job or "modeling-the-expert" training methods. Without a more fully developed basis for quantifying actual training needs, the use of simulators in professional development and marine licensing will continue to be based on perceived needs and professional estimates. To apply instructional design, it is necessary to have detailed, relevant task and subtask analyses.[1] As noted in Chapter 1, although

---

[1] The U.S. Navy has used simulation to conduct shiphandling training since 1987. Its training syllabus was created by the instructional system design process, based on task analyses and personnel performance profiles. This program may be a valuable source of information and data for the USCG and others in the development of simulator-based training programs.

there is a significant amount of literature on job-task analysis, much of that material is either dated and needs to be updated or has only limited application to the marine industry and to mariners' training needs. The task analyses needed to define training needs in instructional design should be detailed and include descriptions of steps required to complete identified subtasks.

Applying Job-Task and Performance Analysis

Instructional objectives, curricula, and, to some extent, instructional approaches are designed to satisfy perceived needs and expectations of training sponsors. Perhaps as a result, the instructional quality of the simulator experience at all ship-bridge simulator facilities is generally reported to be good by companies that have supported simulator-based training and mariners who have attended courses using ship-bridge simulators. However, because the committee could not find evidence of programs designed to measure and analyze resulting performance, it could not determine whether these courses are achieving optimal effects in improving mariner preparation and performance.

To determine whether a training program meets its defined objectives, it is necessary to develop a system to measure and analyze resulting performance. A systematic approach to marine professional development necessitates improving understanding of job performance. This improved understanding could be accomplished by expanding existing job-task analyses to include dimensions that are generally missing with respect to behavioral elements and specific steps needed to execute each subtask. It is important to recognize that not all tasks contribute in the same way to overall performance of functions and duties of the job.

There is little current task analysis work in the marine industry. One example of recent work (Sanquist et al., 1994) is a study done by Battelle Seattle Research Center under contract to the U.S. Coast Guard (USCG) to develop a systematic approach for determining the effects of new automation on mariner qualification and training. The program was undertaken in connection with the agency's responsibilities in marine licensing and its goal to "determine the minimum standards of experience, physical ability, and knowledge to qualify individuals for each type of license or seaman's documentation." Although the report, *Cognitive Analysis of Navigation Tasks: A Tool for Training and Assessment and Equipment Design* (Sanquist et al., 1994), is targeted to automated systems aboard ships, the results should provide task analyses at a level that is detailed enough to effectively apply instructional design to relevant training program development.

The following descriptions of the process are summarized from the report.

> To maintain the safety of our waterways, the U.S. Coast Guard needs to assess how a given automated system changes ship-board tasks and the knowledge and skills required of the crew. . . . Four different, but complementary, methods are being developed.

1. *Task analysis.* A task analysis technique has been devised that breaks down a ship-board function, such as collision avoidance, into a sequence of tasks . . . the current approach synthesizes various existing methods and adapts them to the marine operating environment.

2. *Cognitive analysis.* This method looks at the *mental demands* (such as remembering other vessel positions, detecting a new contact on the radar, etc.) placed on the mariner while performing a given task. The cognitive analysis identifies the types of knowledge, skills, and abilities (KSAs) required to perform a task and highlights differences in mental demands as a result of automation.

Cognitive analysis is applied to an *operator function model* (OFM) for task analysis. OFM provides a task analysis that is *independent of the automation,* i.e., OFM defines the (1) tasks, (2) information needed to perform the tasks, and (3) decisions that direct the sequence of tasks, regardless of whether they are performed by the mariner or the equipment.

3. *Skills assessment.* This method evaluates the impact of automation. It takes the results of the task and cognitive analyses and determines the types of training required to instill the needed KSAs for performing the ship-board tasks. *These will be compared with current training courses to highlight any new training needs.*

4. *Comprehensive assessment.* This method addresses the large number of problems resulting from an operator's misunderstanding of the capabilities and limitations of an automated system. For example, when the radar signal-to-noise ratio is poor, the ARPA [automatic radar plotting aids] may "swap" the labels of adjacent targets. If the mariner is not aware of this limitation, he may be navigating under false assumptions about the position of neighboring vessels, increasing the chances of a casualty. Comprehensive assessment will identify misconceptions about automated systems that could then be remedied through training or equipment redesign.

The USCG intends to use this methodology to highlight necessary training and licensing changes. In licensing, for example, the agency's analysis found that use of automatic radar plotting aids in performing the collision avoidance function eliminated nearly all computational requirements on the mariner. Yet application of the cognitive analysis technique to a sample set of questions from the radar observer certification test found that approximately 75 percent of the questions tested computational skills. Given these results, the report concluded "it would appear that there needs to be a shift in emphasis from computational to interpretive questions on the radar observer certification exam."

DETERMINING TRAINING METHODS

Selecting the Training Media

Mere possession of a ship simulator or other training device and the presence of licensed mariners as instructors do not guarantee the effective and credible use of simulation or effective learning. The presence of specialized features, such as physical motion platforms and high-fidelity graphic images, do not in themselves guarantee a relevant and meaningful training experience. To be effective, simulator resources must be matched to instructional objectives. If simulation is relevant to training objectives, then the type (see discussion in Chapter 2) and level (see discussion in Chapter 4) of simulation, including fidelity requirements, need to be determined.

The way a simulator is treated is important in creating a perception of reality among trainees. If the instructional staff treats a simulator as if it were a ship and the simulator environment as real operating conditions, then the trainees are more likely to treat the experience as "real." In creating the illusion of reality for a limited-task or higher-level simulator, attention needs to be given to accuracy requirements for the mathematical models that drive the simulation (Appendix D). Creation of the training environment is discussed in greater detail in Chapter 4.

No quantitative research was identified by the committee that would establish the relative merits of different approaches to marine training and the types of training offered. There is, however, a research basis that supports the application of different levels of simulation to achieve certain training objectives for cadets, mates, masters, and pilots (Hammell et al., 1980, 1981a, 1981b, 1985; Gynther et al., 1982a, 1982b, 1985). The guides from this research are either preliminary or dated as a result of recent changes in manning and automation. Nevertheless, the guides remain a principal reference and could be used as a starting point for instructional design. The committee observed that few, if any, facilities appear to be using these materials for this purpose.

Mariner instructors reported to the committee that the degree to which a participant is familiar with the training media affects the media's relative value with respect to individual learning. Media familiarity also influences individual performance during a simulation. Just as in real life, a mariner becomes more confident in his or her operating performance of a ship-bridge simulator, manned-model, or radar simulator as his or her familiarity with operating characteristics and specific operating conditions increases. The instructional and training value of all marine simulator-based training media are also affected by the nature and form of instruction, including the operational training scenarios (see discussion of types of scenarios in Chapter 4.)

Defining the Training Program

Curricula

A complete ship-bridge simulator-based training course curriculum will typically include information on the following:

- overall course objectives,
- characterization of course participants and numbers,
- characterization of participant educational and professional backgrounds,
- course structure,
- course timetable,
- individual simulation exercise planning sessions,
- individual simulator exercises and their content,
- individual exercise objectives,
- instructions to staff on the methodology of simulation exercises and subsequent debriefings, and
- method of evaluating participants, as applicable.

Appendix E contains outlines of two simulator-based training courses.

Exercise Scenarios

Scenario Design. Once a simulator-based training program and its objectives have been defined, exercise scenarios should be developed. The following factors should be taken into account when designing these scenarios:

- type of simulator (e.g., special task, full mission);
- geographical database;
- mathematical model ship type and, to the degree relevant to training objectives, the model's fidelity with respect to ship maneuverability in restricted shallow water with small underkeel clearances;
- type and structure of exercise scenario required to achieve the exercise objectives;
- exercise length;
- method of briefing and debriefing;
- cost effectiveness;
- level of fidelity and accuracy needed to support training objectives (e.g., quality, field of view, cues in the visual scene, and accuracy of trajectory prediction); and
- validation requirements.

Scenario creation is crucial to optimizing the training value of individual exercises. Simply creating a realistic scenario does not necessarily result in operating conditions that will evoke desired student responses, create an effective illusion of reality, or create real-life pressures (Edmonds, 1994).

Developing situations intended to challenge or test trainees is sometimes accomplished through scenarios involving role playing. In one possible situation, assignments could be reversed, with seniors placed in subordinate positions and junior personnel in senior positions. The objective is to create a pressure situation in which it becomes apparent to participants that improved interpersonal dynamics and communications are needed to reduce the potential for human and organizational error. This form of role playing seems to work. It should, however, be very carefully debriefed to avoid any negative impact on the confidence of junior personnel whose performance may have resulted in a failed solution during the exercise.

In some bridge team management training, positional shifts in roles can be used to show junior officers the tasks of a master and to remind senior deck officers of the difficulties and limitations of watchkeeping. In the case of pilot training, the pilot and master roles can be reversed to increase the pilot's sensitivity to the master's concerns when the pilot is controlling ship movement.

It also is possible to stimulate lessons through more subtle, yet equally effective, means. For example, a delegation from the committee participated in a simulation involving a crew change, a watch relief, two ports unfamiliar to the new watch officers, and a transit speed that was excessive for the situation but not readily apparent. As the scenario unfolded, bridge team members created enough pressures and problems for themselves without any assistance from the instructional staff. The need for more effective passage planning and improved communications among bridge team members was no less apparent than it might have been in a situation artificially influenced by role reversals or problems inserted by the instructor.

Scenario Validation. Once designed, an exercise scenario must be validated. Validation is necessary to avoid variations in the scenarios that could adversely affect training objectives or provide inaccurate information or insights and therefore contribute to human error during real operations. Care must be taken to ensure that relevant cues are present. In cases where individuals are being prepared for shiphandling and piloting on specific waterways or vessels, higher levels of visual scene fidelity and trajectory prediction accuracy are indicated. These factors are especially important in operating conditions involving restricted shallow water with small underkeel clearances.

It can be an advantage for instructors to visit and familiarize themselves with the real geographical area they are simulating. A visit and local knowledge also help instructors incorporate appropriate visual cues and local operating procedures, particularly if the instructor may have to play the role of a local pilot. Alternatively, a mariner with prior experience in the simulated area could serve as a design consultant.

The instructor must be satisfied that the exercise can be concluded in a way that is relevant to exercise objectives and that the scenario can achieve exercise objectives enroute. The operational result may be successful or unsuccessful, as long as training objectives are satisfied. Only then can the scenario be used with confidence to effectively satisfy training objectives.

There is no standard methodology for validating exercise scenarios. The instructor or instructional staff generally perform this function subjectively, sometimes in coordination and consultation with representatives of organizations sponsoring the training.

Use of a Control and Monitoring Station

Once an exercise is validated and is part of the training course, it must be correctly controlled within the objective parameters. During an exercise, the instructor has to control and monitor environmental conditions and other vessel traffic and initiate or respond to internal and external communications. Depending on the complexity of the scenario, the instructor may also directly control pilot boats and tugs to initiate equipment failures, or he or she may delegate some control to the simulator operator. More important, the instructor must monitor trainee actions and performance, noting mistakes, omissions, and any other relevant information for subsequent discussion and analysis at the debriefing.

Exercises using a ship-bridge simulator (limited task to full mission) are about navigating and handling a ship—not about driving a simulator. The more the trainee "thinks ship," the more he or she will benefit. The instructor's actions must be aimed at making the training environment as realistic as possible. The training process may be helped if the instructor feels like he or she is aboard ship, not just in a simulator control station. This situation is particularly important if the instructor also performs a dual role as master in the bridge watchkeeping course for cadets and junior officers.

When an exercise commences, the vessel's passage and safety are in the hands of the master and bridge team and will be successfully or unsuccessfully concluded by their efforts alone. The instructor is often not present on the bridge during a normal exercise. The exception is when trainees will benefit more by the instructor's presence as an advisor or teacher.

While a simulator exercise is running, the instructor must ensure that the exercise objectives are followed. Input must be consistent with the course objectives. It is all too easy for an inexperienced instructor to "inject some excitement" into the proceedings. This action may destroy planned objectives and, at worst, cause trainees to believe they are in a "simulator versus the students" scenario. If this occurs, all realism and training effectiveness will be lost. Incidents must not be input unless they are planned. In any case, trainees cause occasional incidents themselves through human error.

Complex scenarios, particularly those involving shiphandling using tugs, may so overload the instructor that effective performance monitoring of trainees is precluded. In such situations, two or more instructors may be needed to run the exercise, each with specifically allocated functions.

Effectively observing students in a simulated maritime environment is a skill in itself and is central to conducting an effective debriefing. The seagoing experience of a senior mariner is essential to this role and further emphasizes the need for highly qualified and motivated instructors and assessors.

Duration of the Training Program

The committee found no studies of the optimum length of simulator training time or of the optimum balance among lecture, simulator operation, and review of performance in maritime training. Conceptually, the duration of the course needs to be synchronized with the curricula and learning patterns to support overall training objectives. This approach may or may not be cost effective. Most existing simulator-based training courses last between one and two working weeks. This decision may be due as much to commercial constraints of existing simulator-based training as it is to requirements of the training itself. Other factors that may affect simulator training time include:

- the relatively high front-end cost of intensive simulator-based training compared with the low direct cost of on-the-job learning,
- shipping companies have limited resources available for training, and
- some prospective trainees may be unwilling to devote personal time to training.

Short courses typically compress course content, which may be a disadvantage from the perspective of learning and transfer effectiveness. Although mariners can be exposed to training scenarios that might take years to experience during actual operations, compressed courses provide little opportunity to contemplate results of individual training sessions.

This lack of time to reflect may be especially significant for individuals who have limited nautical experience, such as cadets, or are unaccustomed to simulator-based training media. Conducting training with the same content and actual training time but over a somewhat longer period, provides time for students to contemplate results and plan for subsequent training.

An alternative might be to divide course content equivalent to five days of actual training time into training modules performed one day per week over five weeks or more. This approach has been used successfully at the U.S. Merchant Marine Academy to progressively improve cadet bridge team and watchkeeping performance (Appendix F).

Debriefing Techniques

The final and particularly important part of each training session is the debriefing, which takes place once the simulator exercise is concluded, successfully or unsuccessfully. At this point in the program, lessons of the actual exercise are reinforced, and the trainee is reminded of the exercise objectives and informed of any additional objectives that were not previously divulged.

The simulator can be an effective tool in the debriefing. The ability to record and play back a scenario and to analyze the actions and judgments of the bridge team can assist in assessing team and individual performance.

The debriefing can be led by the instructor or by the trainees themselves (usually with an appointed observer leading), with the instructor acting as a "facilitator." A debriefing of a bridge watchstanding course for cadets or junior officers, for example, is better led by the instructor, whose experience and firm hand will keep the session "on course." A bridge team management course, on the other hand, would probably benefit from the facilitator approach because of the general experience and interactions of a given group. Deciding which method to use should be based on the trainee's level, experience, cultural and ethnic background, and the course type. Language limitations may also have to be taken into account.

To apply the debriefing method with a facilitator, one or more trainees are delegated prior to the simulator exercise to observe the actions of their colleagues throughout the exercise. This observer will open the debriefing by critically examining and commenting on two questions: what went right and what could be improved?

The role of the facilitator or instructor is to allow students, through their discussions, to discover why some things went right and others went wrong. The observer's comments can be recorded for further discussion by group members. The facilitator must be free to criticize and focus attention on lessons learned. Each member of the student team may then be asked to comment before the facilitator or instructor summarizes the session. A decision follows and conclusions are drawn.

Using such a technique for debriefing means that the instructor provides little direct advice during the session. Criticisms are made by group peers and are thus often more readily accepted (even between junior and senior officers). Trainees control the discussion and maintain their own defined reference boundaries. This method encourages trainees to draw their own conclusions and assess their own performance, strengths, and weaknesses. Experience has demonstrated that this approach is most effective if debriefing rules are established at the beginning of the course.

The relationship between instructor and trainees should be a relationship between professionals. Debriefings are vital. If they are too short, unimaginative, or instructor-dominated, little will be gained from the exercise. Trainees must have the liberty to express misgivings and admit failure without fear of penalty or ridicule. They should be encouraged to perceive where they could have performed better to learn from the experience.

Advantages of group discussion include:

- discussion stimulates critical thought,
- trainees learn to substantiate their statements, and
- trainees learn to systematize their thoughts.

During group discussions, the facilitator must avoid:

- discussions that become too time consuming,
- misdirection of group discussions,
- session domination by a few trainees,
- social tensions, and
- animosity among participants.

TRANSFER AND RETENTION OF TRAINING

Transferability of Simulator Training

As discussed in Chapter 2, use of simulators for training is based on the assumption of transfer (i.e., skills and knowledge learned in the classroom can be applied effectively to relevant situations outside the classroom). One unresolved question is a *quantitative* assessment of the transferability of simulator training to the real world. Historically, training effectiveness evaluations of simulators have been developed within the commercial air carrier industry. Studies undertaken in the late 1940s by Williams and associates (Flexman et al., 1972) established the effectiveness of flight simulators for training pilots to fly light, single-engine aircraft. The methodologies developed for these evaluations have been used to demonstrate the effectiveness of simulators for the instruction of a variety of flight skills (Povenmire and Roscoe, 1971, 1973; Waag, 1981; Lintern et al., 1989, 1990).

These methodologies have been adapted to assess training effectiveness from specific simulator features. While some studies support the notion that higher levels of fidelity add to training effectiveness (Lintern et al., 1987; 1990; Hays and Singer, 1989), others do not (Waag, 1981; Hays and Singer, 1989; Lintern et al., 1989; Lintern and Koonce, 1992). For example, research has failed to support the belief in the commercial air carrier community that motion systems add to the training effectiveness of a simulator. Despite the widespread acceptance of motion systems, the scientific evidence is inconsistent.

It is difficult to determine the validity and degree of equivalency between simulator training and shipboard experience without an evaluation of transfer. The issue is whether it can be determined that skills learned in a simulator can be employed aboard ship. The most systematic way to test the application of this training to shipboard performance would be to systematically compare shipboard performance of simulator-trained individuals (as a group) to performance of a group whose only difference is the lack of simulator training. Logistically, these studies are difficult to execute within the commercial air carrier industry and may be even more difficult to execute in the marine industry, which lacks systematic organizational structure.

Available research generally supports the proposition that there is a meaningful transfer of knowledge and skills developed through simulator-based training to actual operations (Kayten et al., 1982; Multer et al., 1983; D'Amico et al., 1985; Hammell et al., 1985; Miller et al., 1985; O'Hara and Saxe, 1985; Froese, 1988; Douwsma, 1993). If there is a concern, it is the lack of reinforcement of newly learned skills in the traditional workplace. Failure to reinforce skills on board ship is a contributing cause in the failure to transfer knowledge from simulators to real life.

Evidence of Transfer Effectiveness

Systematic Studies

An early study conducted at the Computer Aided Operations Research Facility (CAORF), Kings Point, New York, reported that on entering a particular harbor, students who had received simulator training significantly outperformed students with the same background and experience, but with no simulator training (Miller et al., 1985). However, the methodology employed was elementary, and the results are not conclusive.

Appendix F includes a committee-developed case study of the U.S. Merchant Marine Academy cadet watchkeeping course that has used the CAORF simulator since the early 1980s. The case study strongly indicates that watchstanding knowledge, skills, and abilities can be significantly improved using marine simulation, and that this training carries over to watchstanding aboard ship.

Anecdotal Evidence

Simulators are used in a growing number of training programs. In addition to their longstanding use at some maritime academies,[2] and a number of private and union facilities, simulators have been widely used in the commercial air carrier industry, are increasingly used in the nuclear power industry, and are used in medical training and a variety of other areas.

Since objective evaluation of training effectiveness for any specific use is the exception rather than the rule, the committee believes that widespread use of simulators for training and the accompanying belief in their effectiveness constitutes anecdotal evidence of training effectiveness. Indeed, one reason offered for the steady improvement in airline safety since the 1970s has been use of advanced simulators to train pilots for situations too dangerous to try in the air. In

---

[2]Ship-bridge simulators have been used for cadet training at the U.S. Merchant Marine Academy since the early 1980s. Ship-bridge simulators have only recently been installed at the state maritime academies in the United States.

the commercial air carrier industry, for example, it is believed that simulator-based training has considerable value. This belief is bolstered by the observation that airline pilots who transition into a new role in the cockpit via simulator-based training (with no formal in-the-air training for that role) are competent.

The general opinion of mariners who have taken simulator-based courses and the shipping companies who sponsored them is that those courses *are* effective, if not optimal. Shipping companies are using simulators more frequently. In the absence of requirements, they would not be doing so if they thought the training was not cost effective.

Some of the lessons in these training courses, however, may not completely or uniformly be applied in the real world. Learning transfer may fall short because shipboard organization and operating practices have not, in many instances, been restructured to facilitate the introduction and use of these concepts.

MEASURING TRAINING PROGRAM EFFECTIVENESS

The Need for Training Program Evaluation

Anecdotal evidence can, however, be suspect, and apparent effectiveness based on usage patterns and opinions can be misleading. Successful on-the-job performance of those who have undergone simulator-based training could be due to factors unrelated to the training itself.

One element of the instructional design process is continual analysis and improvement of the training program. There is always a concern about the effectiveness of a new or even existing training program. Essentially, the issue is whether trainees learn what is necessary for on-the-job performance. Belief in the training effectiveness is generally based on whether trainees pass the course and perform successfully on the job. There are, however, a number of uncertainties in this sort of evaluation. There is, for example, the question of short-term versus long-term effects, an issue of particular concern for intensive courses of short duration.

In cases where evaluations are conducted within the structure of a training program, the course material may be only marginally relevant. Recent evidence suggests that many technical training programs teach marginally relevant skills, and graduates of those courses have to be retrained when they are placed in an operational environment (Lave and Wenger, 1991). It is not uncommon to hear complaints that new graduates from a training course do not have the necessary operational skills and must be retrained on the job (Lave and Wenger, 1991; Hutchins, 1992). Thus, even when graduates are successful, it is possible that they acquired essential skills on the job, as is typically the case in marine operations.

Satisfactory performance on an examination within a course structure does not ensure that the training was effective. Tests are invariably oriented toward the material taught in the course and may be no more relevant than the course itself. In addition, formal testing may fail to capture subtle but critical aspects of operational skills. Typically, formal tests examine those aspects that are easy to frame and evaluate by standard grading methods. In complex and diverse tasks, formal testing rarely succeeds in evaluating the depth of knowledge and skill needed for operational performance of a multiplicity of tasks while under stress.

Strategy for Evaluating Training Programs

Training program effectiveness should be evaluated as a part of the instructional design process. There are systems for evaluating programs, but many of these are flawed or are not properly applied. The committee developed the following procedures to guide training program evaluation.

A strategy to evaluate the effectiveness of training programs should be relevant, comprehensive, and consistent. To be *relevant*, the strategy should assess skills that are central to the job, especially those that are difficult to teach and difficult to acquire. To be *comprehensive*, the strategy must permit evaluation of the quality of all essential skills and detection of critical omissions. One method is to evaluate the performance of individuals who have completed the course. The aim would be to ensure that ongoing programs realize essential goals and to evaluate whether modifying the training would result in desired enhancements for on-the-job performance both immediately following training and in the long term (NRC, 1991).

These goals might best be met by assigning experienced practitioners to evaluate bridge performance. These evaluators should be carefully selected and trained, remain independent of the conduct of training (to avoid "ownership" in the training product or interpersonal relationships that could influence their evaluations), and be experienced enough to qualitatively judge the effectiveness of operational performance. In addition, they must remain familiar with current practice and, ideally, should periodically cycle through line operations, as is required by most commercial air carrier operators.

It is probably not desirable for the evaluators to be totally independent from organizational goals and policy. They must perform their duties according to shared goals and values established by management. They would also, however, have higher-level goals, such as safety and production. The assumption is that experienced practitioners can recognize how well such goals are being satisfied and are sensitive to tradeoffs that are sometimes essential in the pursuit of diverse goals. Nevertheless, evaluators need to be advised of organizational goals so they will consistently evaluate and logically communicate deficiencies to management and the training department.

The evaluation process would need to be minimally intrusive; it should not markedly change behavior from what it would be in the absence of the ongoing evaluation. Although most individuals perform more conscientiously under evaluation conditions, it would be impractical (and probably unethical) to conceal from a bridge team that they are being evaluated. A poorly trained crew is, however, unlikely to be able to perform at a high standard only while they are being evaluated.

The training program evaluation strategy outlined above corresponds in some respects to procedures currently in use at some simulator facilities. Nevertheless, it should be emphasized that this strategy incorporates several key features, the omission of any one of which will jeopardize the process. Evaluation within a course lacks the required independence.

Positive but unsolicited testimony about the quality of training from those who supervise graduates or from graduates themselves are not sufficiently systematic and may not be very reliable. Although statutory authorities sometimes have an on-the-job evaluation process, their evaluators are rarely as experienced with the actual job as is desirable, are rarely current in practice, and are not necessarily sensitive to all competing work goals.

The difficulty of achieving comprehensive and operationally relevant evaluation of those who have graduated from training programs should not be minimized. However, these features are essential if training programs are to be evaluated effectively. The results of training program evaluations should be used to make periodic program improvements.

SIMULATOR-BASED TRAINING INSTRUCTORS

The role and qualification of marine simulator instructors evokes considerable discussion and debate. Some people in the marine simulator field believe the instructor is the most important training element; others believe the trainee is the most important part of the simulation because beneficial changes in trainee behavior and performance are the desired product. A third view is that the simulator and the simulation produced are particularly important.

The view taken in this report is that although all design components are important to an effective course of instruction, the relative quality of simulator-based training depends more on the instructor's capabilities than those of the simulator or the role of the trainee. The instructor is of primary importance because it is the instructor's role to ensure that *all* of the instructional objectives are met.

In developing a training program, the instructional design process requires consideration of the following factors:

- curricula requirements;
- instructor recruitment or selection to meet curricula requirements;
- instructor professional credentials and their maintenance;
- instructional capabilities, including their development and maintenance; and
- instructor capability to operate simulator resources and integrate them into effective learning programs.

Instructor Knowledge and Expertise

As a practical matter, the instructor's subject-matter expertise is essential to instructional design. The instructor's *tasks*, however, are multifaceted. Many of the instructor's tasks (Box 3-3) are in addition to and lie outside of the mariner's nautical expertise.

There are limited opportunities for mariners to undertake instructional roles aboard modern ships. One possible exception is in piloting, where apprentices are developed under the supervision of marine pilots as a normal practice (NRC, 1994).[3] Although some mariners develop good instructional capabilities during their seagoing service, effective application of the instructional design process requires specialized skills that must be developed or refined separately through specialized programs.

The need for specialized skills is even more important in the use of ship-bridge simulators. As an instructional tool, simulation has evolved to a level of technical and instructional sophistication that often requires multidisciplinary expertise and technical support. In such cases, the instructor needs to be capable of working as a member or leader of an instruction team.

The Instructor's Role

In conducting training, the instructor lectures, role-plays, and facilitates. He or she is the intermediary for:

- creating a synergism among student, curricula, and simulator; and
- making the simulation believable and meaningful.

Objectivity

Although watchkeeping courses have not been mandatory for certification of officers in charge of a navigation watch, some national agencies (including the USCG) grant remission of sea time after completion of such courses. In these cases, the instructor, by virtue of the instructional role and student evaluations, is involved in the award of a completion certificate. The instructor has either moral or official responsibility, or both, for ensuring that each trainee's performance has been satisfactory. At the same time, care must be exercised to ensure that interpersonal relationships do not influence performance evaluations and that the

---

[3]On-the-job training is also used in the professional development of docking masters and operators of uninspected towing vessels who pilot tug and barge flotillas on inland rivers and waterways. The pool of docking masters has not been a source of simulator instructors because simulation has not been used in training for their profession. There are only a few instructors with professional towing backgrounds because simulation has to date been used only to a limited extent in the coastwise towing industry (only one instructor as of 1994).

> **BOX 3-3**
> **Instructional Tasks**
>
> - Development of a thorough professional knowledge of the subject matter (e.g., shiphandling, bridge team management techniques, radar operations).
> - Maintenance of up-to-date knowledge about relevant developments in ship operations (e.g., navigation technology, marine safety policies, and procedures including marine traffic regulations).
> - Development of a thorough knowledge of the functional operation of the simulator and its capabilities and limitations.
> - Development and implementation of training courses, including objectives and, if appropriate, integration of these courses into the total training program.
> - Development of simulation scenarios that best support instructional objectives.
> - Communication with marine industry and piloting professionals regarding requirements and details of training courses (i.e., training needs).
> - Preparation of all necessary course material and equipment.
> - Validation of databases and scenarios.
> - Validation of ship models and production of ship-model maneuvering data.
> - Preparation of incoming courses and coordination of schedules and training strategy with other members of the instructional team.
> - Conduct of courses in a professional manner, using proven and agreed-on teaching methods and skills.
> - Supervision or conduct of debriefings.
> - Preparation and development of trainee evaluation process.

award of credit toward a certification requirement does not unduly influence training program integrity.

Flexibility and Sensitivity

The instructor must be capable of adjusting to trainees' different professional experiences. Experienced trainees are already a professional in their field and should be treated accordingly. In these cases, the instructor's role as a facilitator takes on more importance than it does with less-experienced deck officers or cadets.

Currency

The main demand on the instructor who teaches professional courses is one of stimulating the trainee to *rethink* his or her own performance objectively and constructively. The instructor must not only know the subject matter thoroughly, but be up to date on recent events, such as marine accidents and their proximate and underlying causes and developments in equipment and operating practices worldwide.

Instructor Qualifications

Professional Credentials

There is a strongly held position among maritime instructors that all simulation instructors should possess the highest seagoing qualification awarded by a flag state, which is commonly understood to mean a master's license with no restrictions or an *unlimited* master's license. In principle, the marine license ensures subject-matter expertise and the institutional consideration of nautical credibility.

Some people believe that knowledge of the course content and proficiency in instructional skills are paramount, and that possession of a senior marine license alone guarantees neither relevant nor recent content knowledge nor instructional skills. Instructors without formal instructional skills training are most likely applying instructional knowledge rooted in informal on-the-job and apprenticeship approaches to professional development. Although the insights that accompany this background are important for discerning and conveying the marine operations subtleties, practical experience does not by itself prepare an individual to apply modern learning concepts.

The Need for Guidelines or Standards

Instructor qualification must consider both instructional and institutional factors. From an instructional perspective, the instructor must possess the right *content knowledge* as well as *instructional skills*. From an institutional perspective, the instructor must be credible to trainees and sponsors. In addition, if marine licensing is involved, the appropriate form and level of instructor qualification is also important to the licensing authorities.

The rapid evolution of simulator capabilities, from desktop computer-aided instruction and presentations to full-mission ship-bridge simulator, suggests that there should be more formal standards for qualification of simulator instructors. With a few notable exceptions (discussed below) there are no professional, industry, or national guidelines, standards, or requirements for certifying instructors, either through professional organizations, marine industry, education, training programs, or government agencies. Nor is there a specific professional code of ethics for instructors involved in mariner training. Generally, the instructional capabilities of instructors is determined by employers through job interviews and review of professional credentials.

> **BOX 3-4**
> **Samples of Instructor Training Programs,**
> **Maritime Academy Simulator Committee (MASC):**
> **Draft "Train-the-Instructor" Course**
>
> In pursuit of one of its goals as a committee, MASC has been working to develop standards for simulator instructors. MASC considers it mandatory that all simulator instructors be required to attend a course that covers the following subjects:
>
> - generic laboratory teaching,
> - development of a ship-bridge simulator-based learning system,
> - instructional strategy for simulation via a design workshop,
> - hydrodynamics,
> - debriefing techniques,
> - instructor attributes,
> - exercise design, and
> - grading.
>
> MASC is currently refining the course structure and curricula.

Boxes 3-4, 3-5, and 3-6 summarize the focus of three different "train-the-trainer" programs. Box 3-4 summarizes an effort by the Maritime Academy Simulator Committee to develop a training program for simulator instructors at U.S. merchant marine academies. Box 3-5 gives samples of extensive courses at the Southampton Institute, Warsash Maritime Centre in the United Kingdom, for training instructors who teach on a full-mission ship-bridge simulator. Box 3-6 summarizes a government-required training program in the Netherlands.

In the United States, a de facto certification of instructors occurs through the USCG's administration of course approvals for training programs used, in part or in whole, to satisfy certain federal marine licensing requirements. The agency has established criteria regarding instructor qualifications that must be met to receive course approvals. Evidence of training in instructional techniques is required, and a simulator facility must notify the agency of any changes in instructors, including the credentials of the individual who will be teaching the course in cases where some sea-time equivalency or licenses are awarded.

In December 1994, the USCG began examining an internal proposal to establish a formal certification requirement that it would administer. The proposal envisioned three categories of certification: certified maritime instructor, designated simulator examiner, and designated practical examiner. The proposal also featured a requirement to use licensed mariners or individuals with comparable experience as instructors. The USCG tasked its Merchant Personnel Advisory Committee to review this proposal as an initial step in determining whether to seek implementation authority and resources. The agency's interest in formal

> **BOX 3-5**
> **Samples of Instructor Training Programs**
> **The Southampton Institute, Warsash Maritime Centre, United Kingdom**
> **Full-Mission Ship-Bridge Simulator**
>
> Key elements of the Maritime Simulation Instructor Training Program include the following:
>
> - All candidate instructors are required to have a class one, masters or chief engineer's license and recent sea service.
> - Training is provided in five subsections: (1) full-mission ship-bridge simulator, (2) radar and vessel traffic system simulators, (3) manned-model ship-handling, (4) machinery space simulator, and (5) cargo-handling simulator.
> - For the full-mission bridge simulator there are five phases:
>
> **1. Duration 5 days. Complete the Bridge Team Management (BTM) course as a student.**
>
> **2. Duration 20 days. Understudy senior lecturer. Goals for the student at the end of the period include:** (1) understand operational philosophy of BTM program; (2) be familiar with bridge equipment; (3) continue to attend BTM lectures and casualty workshops; (4) understudy all BTM exercises, including exercise/planning briefings, simulator exercises, and exercise debriefings; (5) be familiar with all course administrative procedures; (6) receive instruction on the remote data station and on database and exercise construction; (7) receive instruction on the ship simulator instructor control station; (8) undertake pilotage duties on the ride of the simulator; and (9) understand all aspects of course administration.

instructor certification is driven by its interest in encouraging quality training programs, concerns over instructor competency, and interest in using training and marine simulation actively within the marine licensing process.

Teaching Methods

Interdisciplinary Instructional Teams

The instructor needs a wide range of maritime skills. Realistically, it is difficult to find all these skills and qualifications in one individual. Additional training can overcome deficiencies, and spreading of instructional skills over the entire staff can enable a training establishment to focus the required specialization to a particular training need.

A full instructional team would consist of subject-matter experts supplemented by individuals with specialized instructional and technical capabilities in

3. **Duration 15 days. Continue to understudy senior lecturer. Goals include:** (1) be familiar with all aspects of exercise control in the instructor control station for the BTM course; (2) conduct all exercise briefings and debriefings for all BTM exercises; (3) control all BTM simulator exercises under supervision in the instructor control room; (4) continue familiarization in database and exercise construction; (5) be familiar with all BTM course lectures and casualty workshops; (6) begin giving selected lectures/workshops under supervision of a senior lecturer; and (7) be familiar with all aspects of course administration and undertake administrative duties.

4. **Duration ongoing through first year.** (1) Independently conduct all BTM simulator exercises, (2) conduct exercise briefings and debriefings, (3) give selected lectures, (4) undertake various administrative tasks, and (5) undertake relevant pilotage duties on the bridge of the simulator. Under supervision (1) complete presentation of all lectures for the BTM course, (2) understand all aspects of the BTM course in all its versions, (3) undertake familiarization with the emergency procedures course conducted on the simulator, (4) undertake familiarization with pilot training courses on the simulator, (5) undertake familiarization with all special courses conducted on the simulator, (6) develop new exercises, and (7) assist in development of new databases.

5. **Duration ongoing updating.** All full-mission ship simulator lecturers are expected to conduct at least one week's updating at sea each year. The Institute educational system allows five weeks "research and scholarly activity" per year for industrial updating, the presentation of papers at conferences, and other allied activities. In addition, all new staff members are required to complete a one-year course (at the rate of one day per week) on training techniques and the educational systems.

the application of simulation and the setup and operation of computer-based simulators and manned models. In practice, the members of a simulator facility's staff are routinely called on at appropriate times during the course of instruction to support simulations through role playing, to observe student performance, and to provide specialized instruction or technical support. This practice satisfies multidisciplinary needs within the limits of the staffs' resident expertise.

Sometimes specialized support is obtained from parties external to the simulator facility. For example, few facilities maintain a hydrodynamicist on staff unless the facility is also involved with channel design. There may be occasions when an expert in a nonmaritime field may have to be brought in to assist on a simulator-based course. The best example is that of bridge resource management training, where psychologists and specialist in human factors and stress and fatigue can contribute greatly to course content and presentation. The importance of highly qualified, trained, and motivated senior mariners as instructors cannot be overstated.

> **BOX 3-6**
> **Samples of Instructor Training Programs**
> **MarineSafety International, Rotterdam**
>
> The Netherlands government requires instructors to complete a formal course of instruction to prepare them for service at the new MarineSafety International Rotterdam simulator facility. Originally developed by FlightSafety International for commercial air carrier simulator instructors, the course was adapted to the marine simulator field. The one-week course was developed by the facility and was based on the parent company's earlier development of flight simulator instructor training. The course consists of 17 hours of classroom instruction in varying formats plus 2 days of training in the use of the facility's simulators. The classroom segment of the course includes lessons on:
>
> - principles of teaching and learning,
> - lesson planning,
> - student-instructor relations,
> - effective communication,
> - oral questioning techniques, and
> - the profession of instructing.
>
> NOTE: FlightSafety International is the parent company of MarineSafety International (MSI), which operates marine simulator facilities at Kings Point, New York, and Newport, Rhode Island. MSI is a partner with the Port of Rotterdam in MarineSafety International Rotterdam.

Lead Instructors

Use of Lead Instructors. Use of lead instructors has been possible because of small class sizes and the individual lesson content. As a practical matter, the content of each exercise or drill that can be effectively overseen by a single instructor more or less coincides with the level of detail and interaction that trainees can accommodate. On the other hand, the use of a single instructor rather than a dedicated, multidisciplinary instructional team is often influenced by cost. If a facility can afford only a single instructor or a small instructional staff, then the emphasis will be on nautical credibility rather than on staff instructional skills and proficiency. These factors may or may not result in optimization of either instruction or learning, depending on all factors present in a given simulation.

Criteria for Ideal Instructor. The committee believes that the ideal lead instructor should have the following skills and qualifications:

- possession of an unlimited master's license or other high-level qualification for specialized training—for example, a license as a marine pilot for pilot training;

- command experience;
- demonstrated effective teaching and communication skills;
- knowledge of the simulator capabilities;
- expert shiphandling skills;
- strong analytical capabilities; and
- current general knowledge of the industry and trainee sector in particular.

Many trainees attending simulator courses are either serving masters in command or senior officers. It is desirable for nautical credibility that the lead instructor's professional qualification be at least the same as the highest qualification for which the trainees are being trained or examined. Perhaps more important, however, the instructor should possess appropriate subject-matter expertise (i.e., if the course is in pilotage, the instructor should be an expert in pilotage). Command experience would be an advantage and is desirable, but is not absolutely necessary.

Many U.S. training establishments provide training for deck officers and vessel operators other than, or in addition to, masters and pilots. For example, some facilities provide training for coastwise tug and barge operations, and one provides training for operators of inland tug and barge flotillas. In these cases, the highest *relevant* mariner qualification is important. In addition to establishing credibility, the instructor and trainees must be able to comfortably relate to each other.

FINDINGS

Summary of Findings

The current approaches to training and professional development in the marine industry are based on a tradition of "modeling-the-expert" and on-the-job training. Many courses that currently use simulation in their curricula have followed the approach of "inserting" the simulation into the training program rather than following a more structured approach to course development.

Systematic application of the instructional design process offers a strong model for structuring new courses and continuously improving existing ones. The primary elements of the instructional design process include:

- determining training needs, including characterizing the trainee population and analyzing job tasks and subtasks;
- determining specific training objectives, including performance measures to determine whether or to what degree the objectives have been met;
- determining training methods to be used, including assessing whether simulation is appropriate to the training objectives;
- developing a detailed course curriculum, including designing exercise scenarios (if simulation is used), determining the duration of the training program, and debriefing techniques; and
- validating the simulator, the simulation, and the curriculum.

Instructional design is an evolving concept. Application of the process should include periodic evaluation of the success or failure of course elements, periodic assessment of the program's overall effectiveness, and regular innovative modifications, as appropriate.

Another issue of concern in the mariner training process is transfer and retention of the training. Use of simulators in training is based on subjective observation and anecdotal evidence that the training system is effective. Very little recent quantitative research has been conducted to determine whether or how effectively simulator training transfers to the work environment.

One of the most critical elements in the application of instructional design is the effectiveness of the instructor. It is the instructor's responsibility to ensure that all training objectives are met. The instructor must possess both content knowledge and instructional skills, especially if he or she is responsible for teaching in a simulator environment. Standards or guidelines defining instructor qualifications are necessary to ensure instructional effectiveness.

Research Needs

In the course of its investigation of the uses of simulators in training and the instructional design process, the committee identified a number of areas where existing research and analysis did not provide sufficient information for the committee to extend its own analysis.

Among the most significant areas identified was the need to update and expand relevant task and subtask analyses for application to the mariner's training needs. For the instructional design process to be effective, the course design should include the definition of training needs based on the steps required to complete identified tasks and subtasks for specific functions. This analysis should include dimensions that have been missing with respect to behavioral elements and specific steps needed to execute each subtask.

This analysis is important for several reasons. First, not all tasks contribute in the same way to overall performance of functions and duties of the job. Second, task analysis is necessary in training course design and performance evaluation. Third, a clear understanding of the skills and abilities required for job performance is necessary for effective performance evaluation (Chapter 5). Fourth, task descriptions and related performance criteria are necessary to design an effective licensing program (Chapter 5).

Other areas for research identified by the committee include:

- the need for a standard methodology for validating exercise scenarios;
- the need for guidelines or standards for qualifying or certifying training instructors;
- research on the optimum sequencing of simulator training;
- the effect of course duration (i.e., short courses that typically compress course content versus courses spread over weeks or months) on learning and transfer effectiveness by different categories of the training population;

- a subset of the study of course duration—this would be an investigation of whether the effects differ in classroom versus simulator-based training for different categories of the training population; and
- a study on whether skills learned in a simulator can be employed aboard a ship. The study might employ a method such as comparing the shipboard performances of simulator-trained individuals to shipboard performances of a similar group with no simulator training.

REFERENCES

D'Amico, A.D., W.C. Miller, and C. Saxe. 1985. A Preliminary Evaluation of Transfer of Simulator Training to the Real World. Report No. CAORF 50-8126-02. Kings Point, New York: Computer Aided Operations Research Facility, National Maritime Research Center.

Douwsma, D.G. 1993. Using frameworks to produce cost-effective simulator training. Pp. 97–101 in MARSIM '93, International Conference on Maritime Simulation and Ship Maneuverability, St. Johns, Newfoundland, Canada, September 26–October 2.

Drown, D.F., and R.M. Mercer. 1995. Applying marine simulation to improve mariner professional development. Pp. 597–608 in Proceedings of Ports '95. New York: American Society of Civil Engineers.

Edmonds, D. 1994. Weighing the pros and cons of simulator training, computer-based training, and computer testing and assessment. Paper presented at IIR International Human Factors in Shipping Week 1994: Strategies for Achieving Effective Maritime Manning and Training, London, England, October 4.

Flexman, R.E., S.N. Roscoe, A.C. Williams, Jr., and B.H. Williges. 1972. Studies in pilot training: the anatomy of transfer. Aviation Research Monographs 2(1). Champaign: Aviation Research Laboratory, University of Illinois.

Froese, J. 1988. Can simulators be used to identify and specify training needs? Fifth International Conference on Maritime Education and Training. Sydney, Nova Scotia, Canada: The International Maritime Lecturers Association.

Gynther, J.W., T.J. Hammell, J.A. Grasso, and V.M. Pittsley. 1982a. Simulators for Mariner Training and Licensing: Functional Specification and Training Program Guidelines for a Maritime Cadet Simulator. Report Nos. CAORF 50-8004-02 and USCG-D-8-83. Kings Point, New York: Computer Aided Operations Research Facility, National Maritime Research Center.

Gynther, J.W., T.J. Hammell, J.A. Grasso, and V.M. Pittsley. 1982b. Simulators for Mariner Training and Licensing: Guidelines for Deck Officer Training Systems. Report Nos. CAORF 50-8004-03 and USCG-D-7-83. Kings Point, New York: Computer Aided Operations Research Facility, National Maritime Research Center.

Gynther, J.W., T.J. Hammell, and V.M. Pittsley. 1985. Guidelines for Simulator-Based Marine Pilot Training Programs. Report Nos. CAORF-50-8313-02 and USCG-D-25-85. Kings Point, New York: Computer Aided Operations Research Facility, National Maritime Research Center.

Hammell, T.J., K.E. Williams, J.A. Grasso, and W. Evans. 1980. Simulators for Mariner Training and Licensing. Phase 1: The Role of Simulators in the Mariner Training and Licensing Process (2 volumes). Report Nos. CAORF 50-7810-01 and USCG-D-12-80. Kings Point, New York: Computer Aided Operations Research Facility, National Maritime Research Center.

Hammell, T.J., J.W. Gynther, J.A. Grasso, and M.E. Gaffney. 1981a. Simulators for Mariner Training and Licensing. Phase 2: Investigation of Simulator-Based Training for Maritime Cadets. Report Nos. CAORF 50-7915-01 and USCG-D-06-82. Kings Point, New York: Computer Aided Operations Research Facility, National Maritime Research Center.

Hammell, T.J., J.W. Gynther, J.A. Grasso, and M.E. Gaffney. 1981b. Simulators for Mariner Training and Licensing. Phase 2: Investigation of Simulator Characteristics for Training Senior Mariners. Report Nos. CAORF 50-7915-02 and USCG-D-08-82. Kings Point, New York: Computer Aided Operations Research Facility, National Maritime Research Center.

Hammell, T.J., J.W. Gynther, and V.M. Pittsley. 1985. Experimental Evaluation of Simulator-Based Training for Marine Pilots. Report No. CAORF 50-8318-03 and USCG-D-26-85. Kings Point, New York: Computer Aided Operations Research Facility, National Maritime Research Center.

Hays, R.T., and M.J. Singer. 1989. Simulation Fidelity in Training System Design: Bridging the Gap Between Reality and Training. New York: Springer-Verlag.

Hutchins, E. 1992. Learning to navigate. In Understanding Practice. S. Chaiklin and J. Lave, eds. New York: Cambridge University Press.

Kayten, P., W.M. Korsoh, W.C. Miller, E.J. Kaufman, K.E. Williams, and T.C. King, Jr. 1982. Assessment of Simulator-Based Training for the Enhancement of Cadet Watch Officer. Kings Point, New York: National Maritime Research Center.

Lave, J., and E. Wenger. 1991. Situated Learning: Legitimate Peripheral Participation. New York: Cambridge University Press.

Lintern, G., and J.M. Koonce. 1992. Visual augmentation and scene detail effects in flight training. International Journal of Aviation Psychology 2:281–301.

Lintern, G., K.E. Thomley-Yates, B.E. Nelson, and S.N. Roscoe. 1987. Content, variety, and augmentation of simulated visual scenes for teaching air-to-ground attack. Human Factors 29(1):45–51.

Lintern, G., D.J. Sheppard, D.L. Parker, K.E. Yates, and M.D. Nolan. 1989. Simulator design and instructional features for air-to-ground attack: a transfer study. Human Factors 31(1):87–99.

Lintern, G., S.N. Roscoe, and J.E. Sivier. 1990. Display principles, control dynamics, and environmental factors in pilot training and transfer. Human Factors 32:299–317.

Miller, W.C., C. Saxe, and A.D. D'Amico. 1985. A Preliminary Evaluation of Transfer of Simulator Training to the Real World. Report No. CAORF 50-8126-02. Kings Point, New York: Computer Aided Operations Research Facility, National Maritime Research Center.

Multer, J., A.D. D'Amico, K. Williams, and C. Saxe. 1983. Efficiency of Simulation in the Acquisition of Shiphandling Knowledge as a Function of Previous Experience. Report No. CAORF 52-8102-02. Kings Point, New York: Computer Aided Operations Research Facility, National Maritime Research Center.

NRC (National Research Council). 1991. In the Mind's Eye: Enhancing Human Performance. D. Druckman and R.A. Bjork, eds. Committee on Techniques for the Enhancement of Human Performance, Commission on Behavioral and Social Sciences and Education. Washington, D.C.: National Academy Press.

NRC (National Research Council). 1994. Minding the Helm: Marine Navigation and Piloting. Committee on Advances in Navigation and Piloting. Marine Board. Washington, D.C.: National Academy Press.

O'Hara, J.M., and C. Saxe. 1985. The Development, Retention, and Retraining of Deck Officer Watchstanding Skills in Maritime Cadets. Report No. CAORF 56-8418-01. Kings Point, New York: Computer Aided Operations Research Facility, National Maritime Research Center.

Povenmire, H.K., and S.N. Roscoe. 1971. An evaluation of ground-based flight trainers in routine primary flight training. Human Factors 15:109–116.

Povenmire, H.K., and S.N. Roscoe. 1973. Incremental transfer effectiveness of a ground-based general aviation trainer. Human Factors 15:534–542.

Sanquist, T.F., J.D. Lee, and A.M. Rothblum. 1994. Cognitive Analysis of Navigation Tasks: A Tool for Training Assessment and Equipment Design. Report USCG-D-19-94. Washington, D.C.: U.S. Department of Transportation.

Waag, W.L. 1981. Training Effectiveness of Visual and Motion Simulation. AFHRL-TR-79-72. Air Force Human Resources Laboratory, Brooks Air Force Base, Texas. Moffett Field, California: NASA Ames Research Center.

4

Matching the Training Environment to Objectives

The instructional design process, when applied to simulator-based training, requires an assessment of the appropriate level of simulation necessary to ensure that training objectives are supported. Included in that assessment should be an analysis of:

- the capabilities of the training platform to produce a sufficiently realistic training environment to meet training objectives, and
- the capabilities of the instructional team to use the resource effectively and efficiently.

How much realism is actually necessary to enable effective learning using simulation has not been scientifically established (Hays and Singer, 1989). Realism, in this context, is defined as having two components—functional fidelity (how it works) and physical fidelity (how it looks).

No recent research was identified by the committee that would quantify how much fidelity is needed for an effective simulator-based training program. There is, however, some agreement that the highest level of fidelity is not necessarily required in all elements of the simulation to achieve quality instruction and training objectives. Early research in this area (Hammell et al., 1980, 1981a, 1981b; Williams et al., 1980, 1982) found that factors other than fidelity were as important, and, in some cases, more important than highest-fidelity simulators in achieving training objectives.

The lack of recent research in the area of simulator fidelity has strongly influenced the perceived need for using the highest levels of fidelity as a defense against training-induced errors in mariner performance. In fact, the highest fidelity is sometimes used even though deliberate departures from fidelity can, in some cases, enhance training effectiveness (Hettinger et al., 1985; Hays and Singer, 1989). The underlying issues are the adequacy of the understanding of human performance and application of education and training principles.

ESTABLISHING LEVELS OF SIMULATION

There is great variability in the physical capabilities of ship-bridge simulators. For example, computer-based, full-mission and multi-task ship-bridge simulators vary by: (1) the technical state of practice when simulators are installed or upgraded; and (2) scale, from small through large ship-bridge configurations. Given this variability, it is useful to think in terms of *levels of simulation* for a given simulator's various physical components (Figure 4-1). Levels of simulation can serve as a technical frame of reference for a subjective assessment of the component's relevance and performance capabilities relative to the training program's instructional objectives.

A simulator may have particularly strong capabilities in some areas and be weak in others. To select the appropriate training platform, it is important to determine which strengths and weaknesses are relevant to training objectives. In some cases, the simulator strengths and weaknesses may not be apparent from a visible inspection or demonstration.

RELATIVE IMPORTANCE OF SIMULATOR COMPONENTS

Each simulator element is important, but their relative importance depends on training objectives. Some equipment is more important to certain functions, duties, or operations than others. Relative importance also depends on the fidelity of details needed to ensure that erroneous or misleading information is not conveyed. Deliberate departures from fidelity need to be clearly understood by all involved. These considerations need to be addressed to avoid the creation of erroneous insights that could potentially affect real-world performance.

The relative importance also depends on the degree of accuracy (including completeness of the visual scene and functionality of the ship bridge and bridge equipment) needed for credibility with trainees. Pilots, for example, typically expect a higher degree of fidelity and realism than do maritime academy cadets.

Perception of realism is another important consideration (Hays and Singer, 1989). The sequence and manner in which simulation components are introduced to trainees can profoundly affect their perception of the training environment as a real-life operating environment. For example, the first ship-bridge simulator component that trainees experience is the bridge or wheelhouse mock-up. To ensure an adequate perception of realism, the mockup needs to be believable so that the trainee and, for that matter, the instructor think "real" ship.

High Level ←————————————————→ Low Level

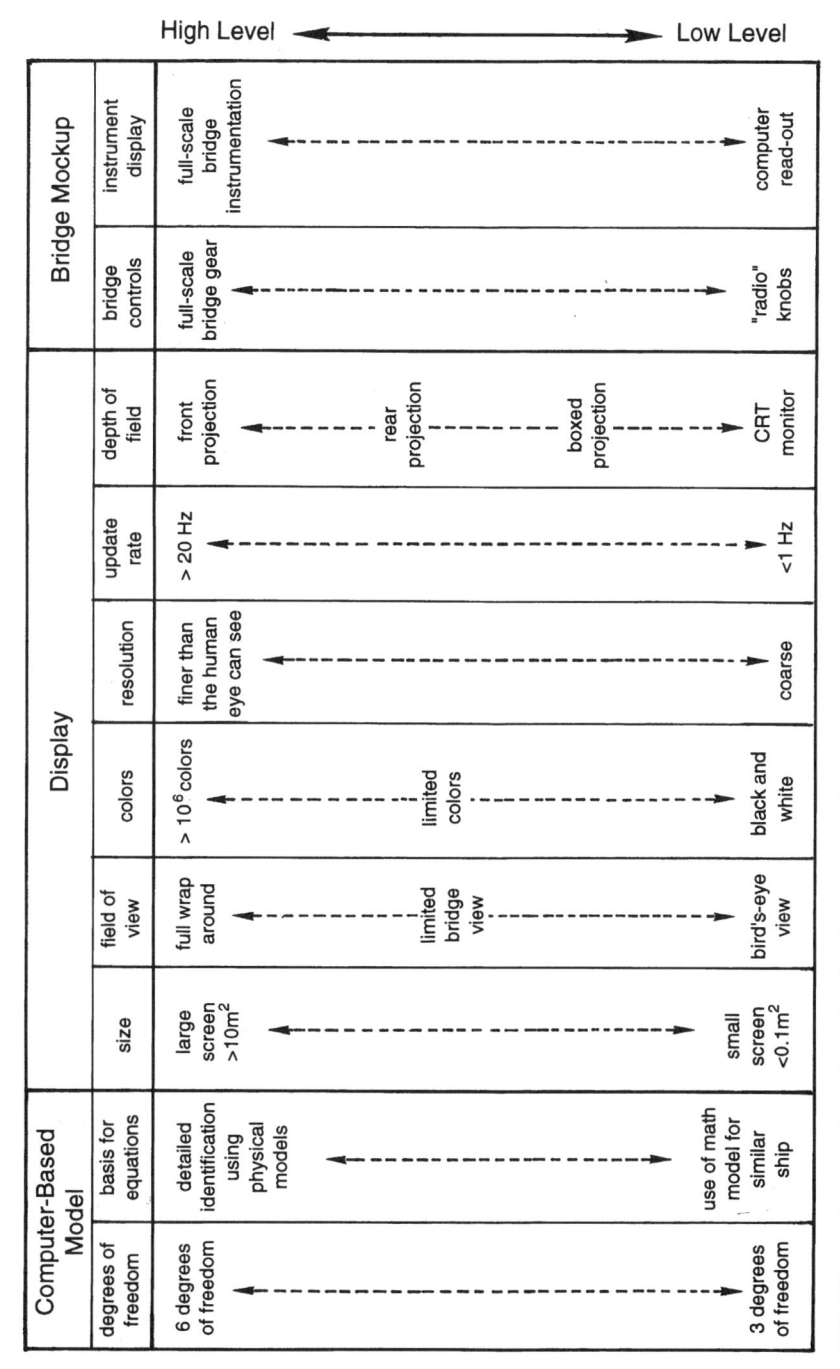

Computer-Based Model		Display							Bridge Mockup	
degrees of freedom	basis for equations	size	field of view	colors	resolution	update rate	depth of field		bridge controls	instrument display
6 degrees of freedom	detailed identification using physical models	large screen >10m²	full wrap around	>10⁶ colors	finer than the human eye can see	> 20 Hz	front projection		full-scale bridge gear	full-scale bridge instrumentation
↕	↕	↕	↕	↕	↕	↕	↕ rear projection		↕	↕
			limited bridge view	limited colors			↕ boxed projection			
3 degrees of freedom	use of math model for similar ship	small screen <0.1m²	bird's-eye view	black and white	coarse	<1 Hz	CRT monitor		"radio" knobs	computer read-out

FIGURE 4-1 Levels of sophistication for simulator physical components.

SIMULATOR COMPONENTS AND TRAINING OBJECTIVES

Creating the Illusion

The simulation illusion begins with how the simulator is treated by the instructional staff. Some instructors always treat the simulator as if it were a real ship's bridge. This approach encourages trainees to accept and treat it as such, or, in the case of a limited-task simulator, as real equipment. Treating the simulation as if it were a real experience is helpful in creating an effective operationally oriented learning atmosphere for mariners.

Bridge and Wheelhouse Mockups

The first physical element of a ship-bridge simulator is usually its entrance. In some facilities, trainees walk from a room onto the bridge of a ship. This approach is not particularly realistic. A number of facilities, however, have been configured to convey a sense that trainees are inside the ship's superstructure, for example, in a passageway leading to the bridge or in a chartroom behind the bridge. Large full-mission ship-bridge simulators, and those with motion platforms, require trainees to climb a ladder to reach the bridge, just as they might aboard a real ship. When trainees walk onto the bridge, they are immediately presented with its structure and configuration. If it does not look like a real ship bridge, trainees are less likely to perceive it as such.

There are other, very practical reasons for a real-life layout. The similarity of the structure and configuration of the bridge or wheelhouse to that of a real ship is fundamental to creating cues that will evoke real-life responses by the trainees. Although not essential, the addition of functional bridge wings could provide valuable cues for conning, shiphandling, applying rules of the road, and particularly for watchkeeping training.

A ship-bridge simulator is different from a visual flight simulator in that it usually has to simulate a library of ship types—from coastwise tugs to very large crude carriers. The bridge configuration generally seems to be more important in creating the perception of reality than its physical size, although there are considerable differences in ship-bridge sizes (and simulators).

The size relationship between a trainee's real ship and the simulator is an appropriate consideration when selecting a facility. There must be sufficient floor space for the comfortable operation of a bridge team. Even old and limited capability (visually and hydrodynamically) ship-bridge simulators can be very realistic with respect to onboard surroundings and can be effectively employed to achieve training objectives.

Small details can also be important to creating favorable mariner perceptions. Although they may have little or no functional utility in a simulation, items such as life jackets, hard hats on hooks, flag lockers, notice boards, and

FIGURE 4-2 Control and monitoring station.

station bills may be important to physical fidelity. The committee found that it is also advantageous if the debriefing area is contained within the "superstructure" of the simulated ship, thereby maintaining the nautical illusion of reality until the debriefing is concluded.

A number of mariner instructors assert that students should not be permitted to visit the control station[1] (Figure 4-2) initially, because premature access may adversely affect their perception of simulator reality. Once trainees' confidence in the fidelity of the simulation has been established, however, showing them the station does not seem to adversely affect their acceptance of the training control environment. Mariner instructors generally believe they need to have the same feel for the exercise as trainees on the bridge; therefore, their control station should be situated and equipped to maintain oversight of the training when they are not role playing or observing on the bridge.

Ship-Bridge Equipment

The relative importance of the ship-bridge equipment for navigation and piloting varies by the functions, tasks, and training objectives. The equipment must be realistic, again to evoke real-life behavior and enhance the transfer of training to real operations. There are no empirical studies to determine the actual

---

[1] The control and monitoring station of a ship-bridge simulator is where the instructional staff inputs scenario parameters and observes the progress of the exercise.

FIGURE 4-3 Estimate of relative importance of ship-bridge equipment for simulator training. Actual importance depends on operating conditions and scenarios.

Ship-Bridge Equipment	Navigation and Piloting Functions						Notes
Key: ● Essential ○ Will grow in importance ◑ Moderately important ◐ Nice to have ✦ Not applicable or of limited utility	Piloting		Watchkeeping	Navigation	Voice Communications	Rules of the Road	
	Shiphandling	Conning					
Engine Controls	●	◑	◑	○	✦	◑	
Propulsion Indicators (RPM, pitch)	●	●	●	◑	✦	◑	
Speed Log (Doppler)	◑	◑	◑	◑	✦	○	
Auxiliary Propulsion Controls	●	○	○	○	✦	○	bow thruster
Engineroom Alarms	◑	○	◑	○	✦	◑	
Steering Console	●	●	●	●	●	●	
Rudder Angle Indicator	●	●	●	○	✦	●	
Rate of Turn Indicator	◑	◑	○	◑	✦	◑	
Master Gyro Readout	✦	○	○	○	✦	✦	
Bridge Wing Gyro Repeaters	●	●	●	●	✦	●	
Magnetic Compass	○	◑	●	●	✦	○	
Visual Bearing Capability	◑	●	●	●	✦	●	
Automatic Pilot	◑	◑	●	◑	✦	◑	
Nautical Charts	●	●	●	●	✦	◑	
Chart Table	●	●	●	●	✦	◑	
Radar	○	●	●	●	✦	●	Radar essential for restricted visibility and for estimating distances at night
Automatic Radar Plotting Aid	✦	●	●	●	✦	●	
Loran	○	◑	●	●	✦	◑	
Electronic Positioning System	○	◑	●	●	✦	◑	GPS, DGPS
Electronic Charting System/ECDIS	◐	◐	◐	◐	✦	◐	
Depth Indicator	◑	◑	●	●	✦	○	
Wind Speed and Direction Indicator	●	●	○	○	✦	✦	
VHF Radio	○	◑	●	○	●	●	essential for communications with assist tugs
Internal Ship's Communications	◑	◑	●	◑	●	●	
Navigation and Signal Lights Panel	○	○	●	○	✦	◑	
Whistle/Fog Signals	○	◑	●	○	✦	●	essential as backup to VHF radio for signals to assist tugs
Reference Publications	✦	◑	●	●	✦	●	
General and Other Alarms	○	●	●	●	✦	✦	
Station Bill and Ship Placards	●	●	●	●	✦	●	
Clock(s)	●	●	●	●	●	●	

importance of equipment to various functions and tasks, but an estimate of relative importance is shown in Figure 4-3. The estimate was compiled by the committee and reflects a composite view based on the committee's operational experience and insights developed by mariner instructors through the oversight of simulator-based training. Some newer equipment—for example, electronic charting systems, either alone or integrated with real-time precision navigation systems—may not be available on most ships.

It is particularly significant that more equipment is essential for watchkeeping than for any other function. Having additional equipment increases the level of complexity and difficulty for officers of the watch. A well-planned and executed ship-bridge simulator watchkeeping course can be used effectively to prepare candidates to operate and use ship-bridge equipment in a multi-task operating environment.

In most cases, actual equipment is used in the simulation so that all bridge equipment can be physically simulated at high fidelity. Visual and audio presentation of bridge equipment can easily achieve high fidelity. The fidelity of electronic displays is typically high, except for maneuvering behavior and trajectory accuracy, which depend on the mathematical model driving the simulation (see Chapter 7 and Appendix D). Typically, the constraints on achieving high fidelity in the bridge mockup are cost and space rather than technological capability.

Visual Scenes

Trainees experience the visual scene more or less concurrently with their entry to the bridge mockup. To create the perception of reality, it is desirable to have a visual scene projected with the vessel either at anchor, at berth, or underway. Once trainees have adjusted to the bridge configuration, they generally begin to review the visual scene. The first component of the scene that will draw attention is the quality of the projection rather than the visual cues. If the quality is poor, then the simulation may be discredited to some extent.

Next, if the simulated location is known to the trainees, they tend to look for familiar visual references. If these are not present, or are not effectively presented, the simulation is again somewhat discredited. The full impact of visual cues on realism occurs later, during actual simulations, when these cues are used to stimulate trainee performance. Because no simulation can replicate all cues used in real life, many simulator-based training exercises begin with a generic port or approach.

Three Types of Scenarios

In general, scenarios are developed based on one of three types of data—hypothetical, real world, and hybrid.

Hypothetical (generic) Scenario. A hypothetical scenario is not based on real-world data for specific ports, fairways, environmental conditions, and other nautical information. Vessel characteristics are often modified from the prototype to reduce complexity or to demonstrate a particular set of characteristics. For example:

- a vessel's bow or stern thrusters may be turned off or reduced in effectiveness;
- the actual draft may be modified to enter a shallow-water port; or
- normal port communications practices, such as vessel traffic services, may be turned off, used or not used, as appropriate to the instructional objectives.

Generic situations may be used for simulator familiarity and facilitation, for new vessel familiarization, or to demonstrate some particular operating, shiphandling, or resource management problem.

Hypothetical operating scenarios are attractive because the trainee will not be familiar with them. Prior familiarity with a route could mask individual performance weaknesses. The lack of familiarity offered by hypothetical scenarios can be important in ascertaining an individual's ability to react in specific situations—for example, to conduct comprehensive passage planning and to react to situations and conditions encountered along a particular route.

The development of hypothetical data can be tedious, especially if all information normally available from reference publications needs to be replicated to satisfy instructional objectives. Because data are usually made available as an excerpt from the appropriate reference, the ability of a simulation participant to identify and correctly use references will probably not be exercised.

The use of hypothetical data is attractive for marine licensing because standard scenarios and mathematical models could be developed (or specific existing models certified) and distributed for use nationwide. This standardization would provide a consistent technical baseline for performance assessments. For testing purposes, however, it would be appropriate to vary operating scenarios and areas to minimize the opportunity for specialized license-preparation courses, which could reduce the effectiveness of simulation-based performance assessments. Developing a suite of hypothetical operating areas with the necessary supporting reference materials would be resource-intensive.

Real-World Scenario. Operational scenarios based on real-world data can be used to simulate a specific port, fairway, or prevailing operating conditions. The vessels employed need to be as close to the prototype as mathematical modeling and resources allow to avoid conveying false or misleading information about vessel performance in specific settings. Such real-world situations can be used to train masters and vessel teams for new vessels or for passage into a previously unvisited port. Real-world data are particularly useful when familiarizing individuals, such as cadets, who do not have strong, prior nautical experience.

Real-world data offer the advantage of providing the actual reference publications that are used aboard ship, thereby exercising the participants' ability to identify and use them. The potential problem of individual familiarity with a port, a factor that can influence performance and mask weaknesses, can be overcome by choosing a port area where the participants are not likely to have prior service.

For the real-world scenario, the fidelity and accuracy of the simulation are especially important. Except for at-sea operations, the simulation would need to be validated for each port and waterway area used to ensure that the charted reference points and information from the coast pilot or sailing directions are included. The simulation should also be validated for information such as local geographic references used for navigation, tides, currents, and arrangements for picking up pilots. This information is needed for simulation credibility and to ensure that the simulation is sufficiently accurate in the event that participants subsequently visit that port.

Hybrid Scenario. A hybrid operational scenario combines real-world and hypothetical data. The real-world information is the principal basis for the simulation, with hypothetical features added to achieve specific instructional or performance evaluation objectives. This approach enables the use of actual reference publications with little modification to accommodate additional hypothetical features. The simulated vessel might be true to the prototype or modified for training or assessment. Such situations add realism to the scenario, but permit larger variations in instructional design. Specific, actual features of the real-world situation may be included or excluded; the fairway can be made wider or narrower or shallower or deeper; and port operating procedures can be modified to system capability or to instructional objectives.

The use of a hybrid scenario is attractive for evaluation purposes because hypothetical features can be added to a facility's existing suite of operating areas and validated as appropriate to instruction or licensing objectives. The hypothetical features could be periodically changed to reduce the opportunity to compromise performance assessments through license-preparation courses. Conceptually, the marine licensing authority would only need to revalidate the changed features rather than the whole simulation. The Computer Aided Operations Research Facility used this approach when it created Port XYZ in the 1970s. It was a hybrid of New York harbor.

Computer-Generated Images and Depth Perception

Visual scenes are presented as computer-generated images (CGI) (Hammer, 1993) in four principal ways:

- front projection,
- rear projection,
- boxed projectors, and
- cathode ray tube (CRT) monitors.

> **BOX 4-1**
> **Computer-Generated Image (CGI)**
> **Projection Systems**
>
> **Front projection systems** generally consist of projectors mounted on top of or above the ship bridge. CGIs are projected onto external screens up to 40-50 feet away that surround the ship bridge. These systems require a theater environment. In some systems, the greater distances are correlated with the optics of the human eye to facilitate the perception of depth.
>
> **Rear projection systems** project CGI onto screens using projectors mounted behind the screens. These systems require a theater environment. The screens are generally closer to the ship bridge than in-front projections.
>
> **Boxed projector systems** generally consist of projectors and flat screens mounted in boxes close to the window frames in the ship-bridge structure. These systems provide limited depth perception, and images can suffer because of the size.
>
> **Cathode ray tube (CRT) monitor systems** place CRT monitors in the window frames of the structure of the bridge mockup; the monitors are "masked" by the frames. These systems have limited depth perception and image size.
>
> SOURCE: Hammer (1993).

Box 4-1 summarizes the differences among these systems.

In addition to CGI projections, nocturnal scenes can be generated using spotlight generators, an older but still highly effective technology for this purpose. Spotlight generators are used either separately or in conjunction with CGI simulations. Generally, nocturnal simulations have achieved higher visual fidelity than daytime simulations. Each approach has technological and training advantages and disadvantages. Which approach a particular facility uses is more a function of cost and space than technological capability, although technology limitations were a principal consideration in simulators installed before the 1990s.

The manner in which visual scenes are displayed affects depth perception, which in turn can influence how individuals perform in the training environment. One method used to improve the depth perception of the projected image has been to curve the projection screen.

Depth of field can best be conveyed by spotlight generators because lights can be displayed at the proper intensity, size, and location. CGI presentations are affected by lower limits of pixel size. Although front projection CGI systems with distant screens provide more realistic depth perception than other CGI systems, the width of the pixel projection increases with distance. The intensity of lights is also somewhat less realistic with CGI, because the lights need to be bigger than they would appear in real life to present the correct intensity. Cojoined projected CGI and spotlight-generator systems allow better visual accuracy and depth perception (i.e., near lights are larger and brighter than far lights).

When depth perception is poor, some simulation participants appear to rely more heavily on electronic navigation aids than they would in real operations, especially during daylight conditions with unrestricted visibility. When port and waterway features are not readily discernable, trainees will often act as if they were operating at twilight or night. This heavier reliance on electronic navigation aids to determine distances is neither good nor bad; the concern is that it is different from actual practice.

The committee did not identify any research on deviations in individual performance or on the transfer of training caused by variations in image projections. The fact that someone might behave differently in a simulation than in real life because of display-induced depth perception problems is, however, an important consideration with respect to achieving training objectives, evaluating performance, and determining competency.

Visual Scene Quality

Visual scene quality in older simulators varies from high for the ship structure projected to low or medium for the traffic ships, navigation aids, and the background (shoreline, shore structures, and water bodies). With the exception of lights (discussed above), the projection quality for traffic ships, navigation aids, and background visual scene is greatly advanced in the latest generation of simulators.

In the past several years, substantial advances in computational power of computer hardware and digital photographic and mapping capabilities have enabled highly sophisticated and detailed CGIs. These advances were quickly followed by considerable attention to the graphic quality of the visual scene by simulator manufacturers. Various shading and texturing techniques are available for improving the general realism of the scene, as well as depth perception and special effects, such as rain, snow, and fog. The level of fidelity depends on the techniques used. For example, although it is possible to correlate shadows on projected vessel decks with the position of the sun, this feature adds significant computational and correlation requirements. Thus, the general practice has been to apply a fixed shadow adjusted only for lighting conditions.

Advances in computational power have also made it possible to greatly improve the representation of moving traffic ships and sea conditions. In the latest generation, six degrees of freedom of vessel motion can be projected in the visual scene and correlated with physical motion platforms. The relative training benefits and cost effectiveness of adding spotlight-generated lights versus improvements in CGI texturing, shading, and special effects have not been established through research.

Regardless of the sophistication of the visual scene possible in any simulator, the visual scene has only to be "believable." Even relatively primitive scenes have been demonstrated to be believable in training applications. The individual visual quality requirements for each course must be determined, based on the training objectives and participants.

Visual Cues

Visual cues—the content of the visual scene—are among the most critical, technical components of an effective simulation. Each simulator must be considered with respect to its computational and projection capabilities. In general, ship-bridge simulators are technically able to provide sufficient visual cues to achieve most instructional objectives. The important consideration is the inclusion of the right content, especially if actual operating areas are being simulated.

Visual cues include the following (some of these general categories may overlap):

- the vessel's hull, superstructure, deck, and cargo-handling features insofar as they can be seen from the vessel's bridge or wheelhouse;
- background details, including geographical reference points, topographical features, shore structures, shore aids to navigation, and water bodies;
- moving models, including traffic ships, buoys, water surface conditions, and special features that convey such motion as bow waves and stern wakes coordinated with ship size and speed; and
- animations, including lighted aids to navigation, running lights, special lights, flags blowing in the breeze, and physical environmental conditions, such as waves, clouds, and local weather conditions.

High fidelity is easily achieved in basic CGI representations of own ship in the visual scene because the position of the vessel's structure is always fixed with respect to the ship's bridge. Therefore, considerable detail that does not change during the course of the simulation can be projected. This detail facilitates depth perception to some extent. Fixed shading and shadows can be added, but there are tradeoffs between realism and depth perception. The own-ship projection is substantially more complicated if shading and texturing are applied and correlated with lighting conditions and movement in the visual scene. The value added by using such cues to achieve instructional objectives has not been established.

The appropriate background details need to be included regardless of whether a generic or actual port area is being simulated. When actual ports are simulated, inclusion of the right cues is critical for functional and psychological reasons because trainees might subsequently find themselves operating in these areas. Attention to accuracy and completeness, insofar as visual features are relevant, is an important consideration in order to avoid the presentation of insufficient, misleading, or incorrect information. This need exists even if the instructional objectives are not specifically intended to support operations in that operating area.

Sufficient background detail is needed to ensure a close correlation between the training environment and the operating environment being simulated. Moderate completeness, including features and navigational aids presented on nautical charts, is generally sufficient to establish the general context.

Identifying the needed cues to create a simulation of a real operating area can be a problem, particularly if the operating area involves harbor approaches and pilotage waters and is not well known to the individuals who are developing the simulation (or if their experience is not sufficiently recent). Details must be consistent with current editions of the *Sailing Directions* and, for U.S. waters, the *Coast Pilot*. Additional local knowledge not included in these two publications must be researched and included where relevant to the safety of vessel operations. Development of this level of detail may necessitate visits to the port operating areas planned for simulation, photodocumentation, consultations with local marine pilots and other local maritime interests, and validation of the resulting simulation by local area experts. Such activities can be expensive. Failure to undertake this applied research and development, however, could compromise the integrity of the simulation.

Field of View

There are two fields of view—vertical and horizontal. The vertical field of view is important in providing adequate background detail for the operating area being simulated. Vertical field of view is technically constrained by the nature of the bridge mockup and the projection system employed. In general, the view out the window of the ship's bridge needs to be completely filled with the visual scene when the trainee is standing at a traditional position for conning the vessel. For mockups in which it is possible to move to a bridge wing, the vertical height needs to be adequate to ensure that the top of the visual scene does not become a distraction to visual observations and bridge operations. Generally, the vertical field of view is adequate in existing ship-bridge simulators.

The horizontal field of view is very important to shiphandling, navigation, piloting, and the interpretation and application of the rules of the road. The ability to provide an adequate horizontal field of view using a ship-bridge simulator is limited by cost, not technological capability. Even with radar and other electronic navigation systems, shiphandlers rely heavily on visual observations. It is necessary to give very careful attention to functional needs and practices with respect to the horizontal field of view. Its scope has the potential for creating controversy. Many simulators are not capable of providing visual scenes either abaft of the beam or astern. Nevertheless, these simulators are used for a number of training courses. This problem should be considered in the context of training objectives and level of simulation or component requirements.

A field of view from the centerline to abaft of the beam on both sides of the vessel and a view astern are essential for full situational awareness and for interpreting and applying rules of the road to these situations. A view abaft of the beam is also helpful for shiphandling and piloting because mariners estimate their relative speed of advance over the ground in piloting waters by looking to the side, not by looking ahead. Changes abeam are a more reliable indicator of actual speed because the visual scene does not change as rapidly ahead as on the beam.

When fixing the ship's position visually, it is good navigation practice to take visual bearings on reference points and to plot these bearings as lines of position. Lines of position need to be separated horizontally by a sufficient angle to avoid large errors in the indicated position. To obtain an accurate fix, it is necessary to take bearings on objects ahead and off and abaft of the beam.

Although 90 degrees on each bow (180 degrees total) is the minimum required to take beam bearings, in practice this is an insufficient field of view because rapid change in bearings on the beam would quickly obscure the object being used as a reference point. One hundred twenty degrees on each bow may be more desirable to provide abaft of beam coverage. A 240-degree field of view may be needed for watchkeeping training because of situational requirements for applying rules of the road and because good watch relief procedures require the relieving watch officer to completely assess the operational situation. In particular, the officer who is relieving the watch needs to be able to see abaft of the beam to determine overtaking traffic (Hammell, 1981a).

It is important for the watch officer to use both visual and electronic cues to confirm the available information because the accuracy of navigation fixes from varying sources varies among operating areas. There may also be undetected errors in the gyrocompass repeater used for taking visual bearings or in the electronic navigation equipment. Therefore, information from electronic and visual sources complement rather than replace each other. Both need to be appropriately included in the simulation to create a complete training environment.

The committee found that the horizontal field of view should be continuous so that vessels approaching on a constant bearing are not obscured by the pilothouse window partitions. A continuous field of view is common for front and rear projection systems but not always available with projection boxes or CRT monitors.

Various techniques have been used to compensate for horizontal field-of-view limitations. For example, one of the ship-bridge simulators visited by the committee generates the visual scenes through 360 degrees, but is limited to 240 degrees of actual scene projection at any time. The simulator has been configured so that the visual scene projection shifts 90 degrees to the right or left when a bridge wing door is opened, the shift being correlated with the door that is opened. Another facility visited by the committee has two ship-bridge simulators. One is a full 360-degree visual scene and the other is a 240-degree simulator that has the ability to rotate the bridge to allow simulation of bridge wings.

If there are no nearby geographic references or aids to navigation, shiphandlers estimate their slow-speed movement through the water by looking over the side. For example, it is possible to determine whether a vessel is still underway and approximately how fast it is going by looking over the side to see the foam on the water's surface going by. Similarly, pilots gauge when to discontinue astern bells during anchoring evolutions by the position and movement of turbulence at the stern of the ship. Currently, the only simulators providing these effects are manned models. Simulating these effects with CGI would require additional projection systems.

Update Rate

The update rate refers to how quickly the CGIs are refreshed. Because jumpiness in the visual scene detracts from realism, the update rate needs to be fast enough so that objects pass by relatively smoothly. Objects passing close by the vessel have particularly heavy update requirements because their positions relative to the vessel change very rapidly in comparison to distant objects.

Until recently, update rates were usually constant across the entire visual scene. Update rates of 20 Hz or greater were considered high fidelity and required large computers (NRC, 1992). The latest-generation computer hardware has greatly raised computational power. The most recent hardware developments are capable of supporting variable update rates of 60 Hz, and greater on demand, and highly detailed visual scenes. Thus, it is possible to support detailed objects, such as ships, at very fast update rates when these rates are needed, thereby maintaining a relatively even, high level of fidelity. These high levels of fidelity are available in only a few late-generation simulators.

Motion

Vessel motion is fundamental to using simulation to create realistic operating conditions. Movement of the visual scene is used to create an illusion of physical motion. Movement of the training platform is used to create a physical sensation of motion. Sometimes, as in visual flight simulators, both visual and physical motion are used to create a realistic operating environment.

Simulation Sickness

Simulation sickness is a form of motion sickness. It occurs when an individual is exposed to cues or conditions that are out of balance with the operating conditions to which humans are acclimated. To date, simulator sickness has not emerged as a problem in ship-bridge simulators or manned models. Simulator sickness may not be a problem for the category of vessels being simulated due, in part, to the lack of strong acceleration and deceleration forces and the general care given to achieving high fidelity with respect to own-ship vessel motion in the visual scene.

The advent of motion platforms with six degrees of freedom and the ability to include heave, pitch, and roll in the visual scene as well as surge, sway, and yaw have the potential to cause physical discomfort. To mitigate the potential for motion sickness, motion platforms and the visual scene need to be well correlated. When they are not, any potential for simulator sickness may be exacerbated.

Motion and Training Objectives

The relative importance of visual and physical motion requirements depends on the training objectives with respect to:

- establishing credibility with the participants in a simulation,
- degree of fidelity needed to support training objectives and transfer of training, and
- the mitigation of artificial conditions that could lead to physical discomfort.

All of marine simulations related to navigation and piloting require that motion be conveyed in a way that enables participants to:

- determine the vessel's movement in relation to geographic references,
- detect other vessels, and
- determine the relative movement between vessels and other objects.

Encounters and interactions with other vessels are routine, and vessel maneuvering and navigation depends to a considerable extent on movement in relation to other vessels and fixed objects. Actual physical motion is very difficult to correlate with movement of the visual scene in a computer-based simulation. Physical motion, however, is not required to achieve most training objectives.

Degrees of Freedom

Motion is categorized in terms of degrees of freedom. A minimum of three degrees of freedom is required for basic movement in the two-dimensional horizontal plane. In the marine environment and marine simulation, movement occurs in a line parallel to the vessel's centerline (surge), perpendicular to the vessel's centerline (sway), and rotational movement (yaw). The remaining three degrees of freedom are roll, heave, and pitch. Surge, sway, and yaw can be simulated using:

- a visual scene,
- a bird's-eye electronic presentation such as radar or automatic radar

plotting aids (the standard devices) or an electronic charting system with integrated real-time electronic navigation signals, or
- a combination of visual scene and electronic navigation aids.

Surge, sway, and yaw are fundamental to vessel movement and therefore must be incorporated into any marine simulation for it to be realistic.

Virtually all forms of marine simulators, including desktop simulators, are capable of simulating surge, sway, and yaw and are thus capable of providing the three basic degrees of freedom. The ability to provide the basic three degrees of freedom does not, however, guarantee a credible, relevant, or even useful simulation by itself. There are great differences between the training media and their representation of motion. Although both a full-mission ship-bridge simulator and a microcomputer desktop simulator can depict an identical maneuvering situation with three degrees of freedom, the training environment in which the situation is experienced is completely different.

Relevance of Physical Motion

The highest level of simulation would include a motion platform with six degrees of freedom and perfect correlation of physical motion and visual scenes. The experience of mariner instructors, however, is that most training objectives can be obtained without physical motion platforms if motion in the visual scene is of relatively high fidelity. Physical motion cues are not critical to the training objectives for most vessels and most operating environments being simulated. In addition, most physical cues in a harbor or waterway caused by waves or swells at sea would normally fall below the thresholds at which they would be detected by human operators, or they would not appreciably affect individual performance if not experienced.

In general, operators of large vessels during normal operations in ports and waterways do not experience rapid acceleration and deceleration forces. They would, however, experience physical motion cues during collisions, touching the pier when mooring, running hard aground, or deceleration forces during anchoring evolutions. Since these events are terminating factors in a simulation, the absence of physical motion to signal their occurrence is only a weakness, not a fatal flaw, in achieving most instructional objectives. Acceleration and deceleration forces also are more likely to be experienced on small ships.

Although there is little information on the degree to which motion platforms contribute to vessel simulation, they could potentially be applied to specialized training situations, including ice operations (e.g., conditions in which deceleration forces are commonly experienced) and mooring of small resupply vessels alongside offshore platforms. To perform effectively under physical movement conditions, however, it is necessary for the individual to first become competent with the fundamentals required to perform effectively in more difficult operating conditions. For this reason, the opportunity to use motion platforms depends heavily on the trainees' experience and expertise, in addition to instructional objectives.

Introducing wave effects through physical motion has been incorporated into some ship-bridge simulators. Using physical motion to represent wave and other motion effects is not a new concept, although it has been used sparingly. One facility originally installed a motion platform in its nighttime spotlight-generator simulator to aid in the instruction of heavy-weather maneuvering. Only heave, pitch, and roll were used because of difficulty in correlating surge, sway, and yaw with the visual scene. Another simulator facility installed a motion platform for training deck officers in handling supply vessels alongside offshore platforms and for ice-breaking operations. Another facility produces wave effects through the combination of physical motion and CGI. In each case, the projection screens have been located off the ship-bridge motion platform. The only marine simulator that features a motion platform with attached projection screens is the U.S. Navy's air-cushion vehicle simulator in San Diego, California. For that facility, the modeling of motion effects, including the correlation with the visual scene, was reported to have been very complicated and difficult.

Manned Models

Manned models always provide six degrees of freedom because the models are scale versions of real ships, and they operate in a real environment. The principal factor affecting fidelity in manned models is the ability of trainees to adjust to the scaling factors with respect to time and size during training, then interpolate their results to actual operations (common scaling factors can increase velocities and reaction times by as much as ten times as fast as real life).

Computer-Based Simulators

For computer-based simulators, depending on the nature and form of the simulation, motion must be evident in the visual scenes, the electronic navigation displays, or both. In the case of full-mission and multi-task ship-bridge simulators, both visual and electronic navigation displays are necessary because both are essential elements of standard navigation and piloting practices.

The primary electronic navigation display that conveys motion is radar. Information from the radar screen or radar ranges and bearings are plotted on a nautical chart. Other options include data plotted from electronic navigation aids, such as Loran or electronic charting and positioning systems. Electronic charting systems are available in only a small number of ship-bridge simulators.

External Elements and Capabilities

A complete simulation features various elements and capabilities that are external to the ship-bridge mockup. Linking of simulators, traffic ships, aids to navigation, noncharted points of reference, vessel traffic service (VTS) systems, assist tugs, and noise are a few examples.

Linking Simulators

Limited-task radar simulators, which consist of a number of linked bridge simulators, have been used for a number of years. Some newer simulator facilities have more than one full-mission ship-bridge simulator. In these facilities it may be possible to link those full-mission simulators to achieve certain, well-defined training objectives. Linking of this type is used most frequently for courses such as bridge team management, emergency procedures, bridge watchkeeping, and pilot training. Benefits of linking might include:

- ability to train rules of the road with multiple ships (though this may be very expensive),
- ability to simulate communications among several vessels, and
- some limited pilot training, such as channel passing and overtaking situations (though these are more effectively trained in manned models).

The cost of linking full-mission simulators, however, can be substantial. Use of several simulators simultaneously requires extra control and may divert attention from the primary task of monitoring student actions (and achieving objectives). In addition, training objectives and scenarios of two or more groups seldom coincide. The linking may therefore be of limited interest to all groups for only short periods.

Traffic Ships

Training scenarios frequently involve interactions with other vessels, referred to as traffic ships. The degree to which traffic ships and their movement patterns realistically portray local operating conditions affects the credibility of the simulation for individuals who are familiar with the simulated operating area. These conditions need to be as faithful as possible to real conditions to minimize the potential for creating false or misleading perceptions that could adversely affect real operations. Traffic ships need to be preprogrammable for automatic operation as well as manually controllable. Otherwise, the simulation would be very limited in its ability to provide realistic operating conditions or to provide a capability for causing traffic ships to react to own-ship maneuvers.

The traffic ship capability varies widely among simulators. Some of the early simulators have very limited traffic ship capabilities, whereas the newest ship-bridge simulators can generate up to 60 other ships of widely varying types. The actual number of traffic ships used needs to be synchronized with the instructional objectives and within the practical management capabilities of the instructional staff.

Limitations in a simulator's traffic ship capabilities can be compensated to some extent by linking simulators. For example, one simulator facility compensates for limitations in traffic ships by linking three limited-task simulators to its ship-bridge simulator. The tradeoff in linking simulators is some loss of the control over the exercise in return for increased interaction among participants.

Aids to Navigation

Aids to navigation are fundamental to marine operations and need to be faithfully replicated in simulations to achieve functional and physical fidelity. Because they are principally what cadets use as references for navigation and piloting, aids to navigation are especially important to cadets. In contrast, experienced mariners tend to combine use of navigation aids and noncharted reference points (such as natural ranges).

Errors in the positioning, characteristics, or numbering of aids will discredit a simulation because that information is readily available on navigation charts. Generally, aids to navigation are displayed adequately, although there can be problems with jumpiness in the visual scene due to slow update rates. There can also be problems with light intensity and size, which affect visual acquisition of these aids and depth perception.

Noncharted Points of Reference

Credibility of the visual scene for mariners depends largely on the inclusion of commonly used noncharted reference points (e.g., a church steeple or clock tower). In general, the inclusion of noncharted reference points is limited more by the cost of establishing and validating the database for the setting than by the technology. Failure to include these references would discredit the simulation for piloting purposes and demonstrates a lack of attention to essential detail by the simulator facility staff and the provider of the visual scene.

Transit Information and Vessel Traffic Services

Most of a vessel's operations during a routine port entry or departure do not need to consider the entire traffic situation within the port and waterway complex. The simulation only needs to consider the traffic a vessel will encounter along its planned route within the simulation time span, sequentially rather than collectively. In some cases, substantial advance passage planning is required because the intended route and timing of passage may depend on prospective conflicting traffic at chokepoints, tides, currents, and changes in berthing and anchorage assignments. In these cases, that information needs to be planned into the scenario.

The means for providing advance transit information to vessels varies by operating area. In some locations, mariners need to listen closely to the radio because there are no information or advisory services in operation. In some areas, VTS and VTS-like systems that provide navigation and traffic information have been installed. Usually communication with a VTS system is not the main objective of ship-bridge simulator-based training. Therefore, the instruction staff will sometimes role-play via voice radio to provide essential information. If more complete VTS role playing becomes necessary, a substantially improved capability is needed. Realistic VTS simulators are available and can be linked with full-mission or limited-task simulators, although training objectives do not normally coincide to any useful degree.

For realism, background VTS communications are sometimes prerecorded and played on radio speakers on the bridge. Care must be taken that the background information does not involve information relevant to the vessel's operation or become distracting, unless specifically needed to achieve instructional objectives.

There are a few facilities that provide both ship-bridge and VTS simulator-based training. This linked training can serve instructional objectives for both vessel and VTS operators. Another alternative is for a qualified VTS watchkeeper to role-play the VTS function. Additional staff may be required to control tugs used in mooring operations on a full-mission ship-bridge simulator.

Assist Tugs

Assist tugs are needed for shiphandling during mooring operations and for tug escorts. The capability of simulating the forces for assist tugs varies by the mathematical model used and is generally adequate to serve instructional objectives. The physical forces exerted by escort tugs when they come alongside are more difficult to model because the data on physical forces involved in these interactions during passage through piloting waters are limited. Recent simulator-based research into escort tugs for tanker operations in Prince William Sound, Alaska, has contributed to improved knowledge in this area (Gray et al., 1994). It is unknown at this time if the insights gained from this research have been adapted to any ship-bridge simulators.

Noise and Vibration

Noise and vibration are important to marine operations because certain noises provide important, if subtle, cues. Mariners routinely detect major changes through shifts in noise and vibrations. The most important noise is the propulsion system (i.e., engine noise). Changes in engine noise and vibration are normally the first indication of a change in shaft revolutions.

Engine noise and vibrations associated with the propulsion system can be added to a ship-bridge simulator in various ways. For example, a low-frequency speaker can be mounted under the bridge deck of the simulator. The sounds and vibrations that are produced need to be correlated with the RPM indicators.

FINDINGS

Summary of Findings

In the application of simulation to training requirements, it should be recognized that there are differences in the levels of simulator component capabilities. Ship-bridge simulators can usually simulate a variety of ship types—from coastwise tugs to very large crude carriers. Each element of a simulation—the equipment simulated, the visual scene, and the motion—is important, but the relative importance of each element depends on the training objectives defined through application of the instructional design process. The creation of the illusion and the treatment of the simulation environment as being a real ship bridge are important elements in training effectiveness.

Research Needs

In applying simulator components to objectives defined through the instructional design process, the committee identified many areas where existing research and analysis do not provide sufficient information for the committee to extend its own analysis.

The committee could find no quantitative data that address the question of how much realism (fidelity) is actually necessary to enable effective learning using simulators and simulations. Research in this area should look both at functional (how it works) and physical (how it looks) fidelity.

There is little information on the degree to which motion platforms contribute to vessel simulation. The committee agrees that motion platforms could potentially be applied for specialized training situations, but it believes that the usefulness of motion platforms will depend greatly on the trainees' experience and expertise and the accurate correlation of motion and visual cues.

REFERENCES

Gray, D.L., J.H. de Jong, and J.T. Bringloe. 1994. Prince William Sound Disabled Tanker Towing Study, Part 2. Report File No. 9282. Prepared by The Glosten Associates and the Maritime Simulation Centre Netherlands for the Disabled Tanker Towing Study Group, Anchorage, Alaska. Seattle, Washington: The Glosten Associates.

Hammell, T.J., K.E. Williams, J.A. Grasso, and W. Evans. 1980. Simulators for Mariner Training and Licensing. Phase 1: The Role of Simulators in the Mariner Training and Licensing Process (2 volumes). Report Nos. CAORF 50-7810-01 and USCG-D-12-80. Kings Point, New York: Computer Aided Operations Research Facility, National Maritime Research Center.

Hammell, T.J., J.W. Gynther, J.A. Grasso, and M.E. Gaffney. 1981a. Simulators for Mariner Training and Licensing. Phase 2: Investigation of Simulator-Based Training for Maritime Cadets. Report Nos. CAORF 50-7915-01 and USCG-D-06-82. Kings Point, New York: Computer Aided Operations Research Facility, National Maritime Research Center.

Hammell, T.J., J.W. Gynther, J.A. Grasso, and M.E. Gaffney. 1981b. Simulators for Mariner Training and Licensing. Phase 2: Investigation of Simulator Characteristics for Training Senior Mariners. Report Nos. CAORF 50-7915-02 and USCG-D-08-82. Kings Point, New York: Computer Aided Operations Research Facility, National Maritime Research Center.

Hammer, J.A. 1993. Visual system fidelity in marine simulators: a practical guide. Pp. 153–161 in MARSIM '93. International Conference on Maritime Simulation and Ship Maneuverability. St. Johns, Newfoundland, Canada, September 26–October 2.

Hays, R.T., and M.J. Singer. 1989. Simulation Fidelity in Training System Design: Bridging the Gap Between Reality and Training. New York: Springer-Verlag.

Hettinger, L.J., R.S. Kennedy, and K.S. Berbaum. 1985. Tracing the etiology of simulator sickness. In Proceedings of the Conference on Simulators, Norfolk, Virginia, March 3–8.

NRC (National Research Council). 1992. Shiphandling Simulation: Application to Waterway Design. W. Webster, ed., Committee on Shiphandling Simulation, Marine Board. Washington, D.C.: National Academy Press.

Williams, K., J. Goldberg, and A. D'Amico. 1980. Transfer of Training from Low to High Fidelity Simulators. Report No. CAORF 50-7919-02. Kings Point, New York: Computer Aided Operations Research Facility, National Maritime Research Center.

Williams, K., A. D'Amico, J. Goldberg, R. DiNapoli, E. Kaufman, and J. Multer. 1982. Simulators for Mariner Training and Licensing, Phase 3: Performance Standards for Master Level Simulator Training. Report Nos. CAORF 50-8007-02 and USCG-D-15-82. Kings Point, New York: Computer Aided Operations Research Facility, National Maritime Research Center.

5

Performance Evaluation and Licensing Assessment

UNDERSTANDING PERFORMANCE EVALUATION AND ASSESSMENT

Mariner Competence and Proficiency

Knowing what constitutes training, competence, and proficiency is important to understanding the nature and role of mariner performance evaluation and assessment. Training is the systematic development of attitudes, knowledge, and skills required by an individual or a team to perform a given task appropriately. Competence is having adequate knowledge or skills to perform occupational activities, to establish employment or licensing authority standards, as defined by performance criteria. Proficiency is demonstrated ability.

The difference between competence and proficiency is illustrated by the traditional mariner licensing process. Merchant marine officers are tested by a written examination during licensing. This method may demonstrate a level of knowledge, but does not demonstrate sustained ability to perform the task or the job. Simulators may provide a practical method for measuring or testing levels of competence and proficiency and the ability to continue to prioritize tasks. Yet few training courses described for the committee rigorously addressed measuring, evaluating, and assessing individual mariner performance.

Measuring Human Performance

Measuring human performance, either simulator-based or real-performance-based, is often difficult. The visible behavior is only part of the performance:

perception, cognition, physiology, and psychology are buried within the individual. Team performance is even harder to measure. With team performance a number of individuals interact, and the interactions need to accommodate authority levels, role and responsibility levels, verbal and nonverbal communications, and power levels.

Human Factors Aspects of Simulation (NRC, 1985) contains comments and recommendations with respect to performance measurement. The study concluded that the development and application of performance measurement in simulator-based training and research requires that the following elements be considered:

- Operational measures and criteria of overall system effectiveness for representative tasks and operating environments are needed.
- Analysis of the hierarchy of goals and control strategies that operators employ in the performance of real-world tasks is needed.
- Measurement for performance diagnosis needs to be developed.
- Team performance measurement where the contribution of each person is to be defined and measured is difficult.
- Automated instrumentation and performance measurement systems are not possible in all simulations, or at least for certain tasks.
- There is no single source, or even a coherent body of literature, to which practitioners can turn to obtain useful data on performance measurement methods and simulation practice.

Performance measurement must consider three levels of relevance: *concept, understanding,* and *performance. Concept* refers to the knowledge base of the individual, the degree to which the correct analysis and response methods are known. *Understanding* refers to the degree to which the individual is able to accommodate and work around missing or inaccurate information to correctly analyze and respond. *Performance* refers to the physical acts that are observable by instructing or evaluating personnel. Performance observation is clouded further when tasks demand multiple and simultaneous responses, are complex, or are to be measured at points of significant stress (which may include extremis or near-extremis exercises).

Evaluation and Assessment Defined

In some fields the terms evaluation and assessment can be used interchangeably. For this study, the terms are given more narrow definitions. *Evaluation* is applied to the formal or informal review of training exercise results: the input is the training program; the output is the evaluation. In this context, evaluation is an element of the instructional design process. The evaluation can be informal or formal, subjective or objective, or both (see "Forms of Evaluation and Assessment" below).

Assessment is used only in the context of the licensing and certification process. Assessment is the testing of competency against specific standard criteria used for certification or licensing. The input is the formal test of competence against a set of stated, standardized criteria; the output is the assessment, either objective (e.g., multiple-choice test) or subjective (assessor completion of a simulation checklist).

This use of more-narrowly defined terminology extends to the terms *instructors* and *evaluators* in the context of training programs and *assessors* in the context of licensing and certification.

FORMS OF EVALUATION AND ASSESSMENT

Within the simulator environment, performance evaluations for trainees may be informal or formal, subjective or objective, or both. Performance assessments for licensing candidates conducted in a simulator environment are always formal, though they may also be objective or subjective or both.

Informal Evaluations

By far the most common type of evaluation is informal. These evaluations, most of which are implicit, are routinely conducted as an integral part of simulator-based training courses. They are typically conducted on an ad hoc basis and are usually not written. The most common form of informal evaluation is the undocumented debriefing of an exercise by an instructor or instructors. These routine, ad hoc evaluations are used to adjust exercise content and timing and to guide trainees toward achieving planned learning objectives.

Instructors also evaluate the results of training to help improve course content and methodology. They evaluate each student's professional background, experience, attitude, and aptitude to select the most appropriate learning methods and measurements. The instructor may also evaluate the results of each exercise to provide expert critiques of performance activity and/or to facilitate trainee exercise debriefings and conduct peer evaluations of the results.

Trainees also conduct evaluations. They continuously evaluate the degree to which a course is moving them toward meeting their personal or professional development objectives. Similarly, the sponsors of trainees make implicit evaluations on a course's value to their organizational objectives. The results of the course performance may or may not be formally recorded and retained. Generally, formal records are not retained, although there are exceptions, such as when grades need to be assigned to meet baccalaureate requirements.

Private, Informal Evaluations

Sometimes, albeit infrequently, training sponsors or pilot associations have requested that simulator facilities conduct a private evaluation of a specific individual who is scheduled to participate in training. Generally, such requests have been borne of necessity. A company may have received indications of a performance problem and have few, if any, other practical options for determining whether there is a problem that merits corrective action, such as additional specialized training.

No data are available on the practice of private evaluations. As a rule, however, simulator facilities have been reluctant to provide private evaluations of individual performance. This reluctance is caused by concerns about possible adverse effects on the credibility of the simulator facility operator's training programs and possible liability. To the extent that private evaluations have been done, they have often taken the form of individual feedback, almost always without the preparation and retention of formal documentation.

Formal Evaluation and Assessment

Because there has been a general reluctance among operating companies, unions, and operators of marine simulator facilities to formally evaluate the knowledge or performance of licensed mariners, formal evaluations are seldom employed in marine simulation. The committee did, however, find instances where formal performance evaluations were conducted on a simulator. These cases included cadet evaluation using bridge watchkeeping courses at maritime education institutions, an offshore towing deck officer training program, an active watchstander course, and a leadership course (discontinued) sponsored by a major shipping company.

In the first case, cadets attending the U.S. Merchant Marine Academy are required to take a course in watchstanding that uses the Computer Aided Operations Research Facility ship-bridge simulator as the principal training aid. Cadets are required to attain minimum performance standards of watchkeeping, including communications, navigation, change of watch, and bridge team coordination practices. These practices are generic and applicable to all vessels, rather than to operating practices aboard a specific ship or within a specific company. All cadets are required to complete this course to graduate from the Academy. A detailed case study of the course is provided in Appendix F.

The shipping company leadership and team-building course required that each participant meet certain final performance criteria for continuing employment. Those participants who did not achieve satisfactory performance levels were permitted to participate in additional course work and practice to reach a predetermined minimal level of performance. These performance minimums were observed while the student was participating in shiphandling simulation.

In the use of formal simulator assessment for licensing, there are task-specific subjective assessments on mandatory radar observer courses and some other radar courses, such as the use of automatic radar plotting aids. The *Master's Level Proficiency Course* recently approved by the U.S. Coast Guard (USCG) (see Appendix F) and offered at the STAR Center in Dania, Florida, contains both a written (objective) and an assessor-scored simulation assessment (subjective) test.

To be effective, formal evaluations or assessments must have standardized and structured monitoring and must include a critique of individual performance in a range of exercises appropriate to the instructional or licensing objectives or criteria. Formal evaluations or assessments must be consistent in method, timing, and responsibility from class to class or test to test, so that results can be compared and contrasted with a high degree of reliability.

Objective Evaluations or Assessments

An objective evaluation or assessment is not subject to evaluator or assessor bias or observational limitations. To use objective evaluation or assessment the performances must be able to be expressed in "yes" or "no" format. These evaluations could use a checklist or a simulator-embedded assessment. In an objective evaluation, the evaluator would note whether a particular practice took place. Objective evaluation may also permit the evaluator to indicate the quality of the practice. Use of the objective method generally requires that the student accept the evaluator as an equal.

Checklists and Task Lists. Checklists and task lists can be useful for measuring performance objectively. Use of such forms, however, requires that:

- the specific and detailed elements of the performance under consideration are well known,
- these elements have been fully articulated, and
- they are accepted by all relevant groups.

As discussed in Chapter 1, there is research in the maritime industry that includes task analyses. These studies have been made widely available for some time, but their results have seen limited use and are now somewhat dated. The International Maritime Organization (IMO) also has promulgated various booklets, *IMO Practical Test Standards*, that provide checklists for criterion-measured evaluation. The few courses that have included formal evaluations have used checklists or task lists, in some cases combined with subjective (qualitative) evaluations of each task performed.

Simulator Embedded. The current level of simulator sophistication permits the simulator program itself to evaluate the student's performance. Based on ship type, ship loading, hydrodynamic and aerodynamic characteristics, environmental conditions, and instructor input, the simulator can evaluate the degree to which the student met the performance parameters established for the run. The evaluation can be portrayed graphically (and in color) showing own-ship's track, rudder and engine commands, other vessel tracks, and the impact of environmental conditions. The student may be given a copy of the evaluation, and a copy may be stored in the computer memory for future evaluations of other students.

The simulator's sophistication cannot hide the essential form of the evaluation: the evaluator, the computer, or some combination of the two determines what the correct performance is and what each successive student's judgments should be. Since there are many acceptable ways to perform navigational and shiphandling tasks, acceptance of such evaluation by students who are master mariners may be limited.

Performance Playback. Simulator playback capabilities can be important. The ability to make audio, video, and plotter recordings of several measures of performance and behavior is a valuable tool in performance evaluation and assessment. Printouts may be made of situation displays, bird's-eye views, and status displays at different stages of the exercise on different scales. These data can be used during training debriefings for informal evaluations and during licensing assessment as one objective measure of the candidate's performance.

Subjective Evaluation and Assessment

Subjective evaluations and assessments are open to interpretation or bias by any or all involved—the evaluator, assessor, or student. While these methods might take the form of checklists or task lists, the evaluation or assessment also includes the observer's qualitative judgment of the efficacy of the student's performance. In the training environment, subjective evaluation requires that the evaluator be accepted as having superior knowledge of the subject matter.

The typical ship-bridge simulator course uses debriefing as the subjective evaluation form of choice. The instructor, student peers, and students themselves may comment on portions of the run that were well done or were not so well done and ways in which performance could be enhanced. These informal (written or unwritten) evaluations carry substantial weight with students and instructors. Personal practices can be compared, results measured, and alternatives explored in conversation and on the simulator.

Use of subjective evaluations can be effective and, in the absence of scientific performance-based criteria, is currently the primary means for ascertaining whether an individual can effectively apply knowledge in conducting actual operations. For example, a subjective evaluation may be used in determining whether an individual can collect, correlate, and interpret considerable information from multiple sources, make appropriate decisions based on this information and his or her nautical knowledge, and perform multiple tasks in the correct time sequence under the routine pressures of actual operating conditions. Indeed, the functions and tasking just described are exactly what is asked of the officer of the watch and, more important, of third mates from the moment they begin to stand their first underway watch.

In the use of subjective assessment for mariner licensing, the qualifications and credibility of the assessor are especially important. As discussed later in this chapter, the licensing authority responsible for the testing, as well as the candidate being tested, must be assured that the assessor has superior knowledge and can be impartial in conducting the assessment. Assessor qualification is an area where the similarities between use of simulation in the marine and commercial air carrier industries are pronounced. In both industries the qualifications and perceived credibility of instructors, evaluators, and assessors are paramount to the individual being evaluated or assessed.

TRAINING AND EVALUATION WITH SIMULATORS

Although there are notable exceptions, evaluation is not systematically applied nor is the methodology for conducting performance evaluation well developed. Most performance evaluation methodologies that have been attempted rely on adherence to prescribed procedures (e.g., operating certain equipment or using correct radio procedures) or subjective evaluations by experts. These evaluations may or may not be based on detailed task analyses and formal evaluation objectives and criteria. In the use of simulation for evaluation, it is important to validate the evaluation procedures used to ensure that they are evaluating applicable competencies (i.e., competencies needed at sea).

Current Evaluation Methods

Traditionally, the primary methodology for evaluation in a marine training program has been observation and feedback from the instructor (or instructors) to individuals and teams. In these cases, the instructor usually has the responsibility for both the input (e.g., lectures, discussions, demonstrations) and the output (evaluation). Often these two distinct roles are held by the same person.

An alternative method, sometimes used in vessel or bridge team and bridge resource management courses has been to assign to other course participants the role of observer and evaluator. This peer evaluation methodology relieves the instructor of the dual role and permits evaluation of instructional efforts based on the strength of the peer evaluation.

There can be two serious drawbacks to peer evaluation. First, peers may be reluctant to evaluate the performance of their peers. They may feel that such feedback is not in the tradition of the industry and that the ideal of unencumbered, individual (master) decision making is being compromised. The peers may also feel that they are insufficiently schooled in the practices under consideration to make useful feedback comments. (Of course, providing feedback to others may be included in the instructional design, since such a practice forces observers to learn from performers).

The second drawback is that the instructor may not allow peer feedback. In the role of evaluator, the instructor may take the lead, criticize or disagree with peers, not acknowledge alternative perspectives, or in other ways use his or her authority to limit peer performance evaluation.

There are a few instances in which individuals other than instructors have been involved in evaluation and feedback. Some companies have used either senior mariners who are not part of the instructional cadre as observers or evaluators. Other companies have used senior shoreside or nonoperating personnel in this role. In these cases, the observers were usually recognized as having specialized skills, knowledge, or experience that could be appropriately applied to the evaluation process.

Evaluation Criteria and Performance Standards

When using simulators for training, evaluation criteria should be carefully defined. Selected performance standards should reflect at-sea competency requirements. Generally these standards will represent a baseline level of required ability, not a standard of excellence. It may be more difficult to evaluate teamwork than individual performance in an objective, measurable way, especially because vital skills, such as judgment, are not easily evaluated.

Because of the need to ensure that everyone meets or exceeds baseline standards, normative-referenced[1] testing and evaluation appear to have limited value in marine professional development. These evaluations are best suited for cadet training or training situations involving several junior third officers in the same class.

Most training evaluation in the maritime industry appears more suited for criterion-referenced[2] or domain-referenced[3] testing. These methodologies allow the performer to demonstrate both the strengths of individual capabilities and the deficiencies, which may be corrected through practice, coaching, or additional experience. For criterion-referenced evaluation to be fully effective, however, it is important to address evaluation in the context of the total instructional design process. As noted earlier, concerns in applying instructional design include limited usefulness and age of detailed task analyses available within the industry,

---

[1]Normative-referenced testing is applied to a particular group of people at a particular time. The scores of the individuals are ranked, and the results permit comparison of performance among individuals or across a group of individuals assumed to be similar in makeup and knowledge.

[2]Criterion-referenced testing is always based on some clearly predetermined and articulated set of performance levels and conditions. Criterion-referenced testing is more closely associated with minimum competency testing.

[3]Domain-referenced testing may be seen as a subset of criterion-referenced testing. Domain-referenced testing requires development and maintenance of relevant job performance measures before training begins. It examines what the person knows or can do, not how well he or she compares to others being testing.

absence of performance standards, variability in the observing and evaluating processes, and reluctance among some in the industry to fund and employ the required changes.

Concerns about Evaluation with Simulators

Within a training program, the inclusion of a formal evaluation by an instructor could adversely influence the interpersonal dynamics between the instructor and the students, or among the students in a team-oriented course. The trust and rapport between instructional staff and trainees is vital and can be damaged if the students are overly concerned about passing a test.

There are also several reasons why some operating companies, unions, and others may be reluctant to measure or evaluate individual performance using a simulator. First, the mariner who attends a training course at a simulator facility has already met the USCG requirements for licensing. Since the license is already valid—and by extension, the mariner is assumed to be fully knowledgeable and qualified—simulator-based training can be viewed by the trainee as interesting and of some marginal value, but not required. It is possible that many mariners would forego the opportunity to train on a simulator if they were required to pass a nonlicense-required written or performance test.

Second, a substantial number of U.S. simulator facilities are operated by unions or union-related organizations. Training is often a benefit of union membership. In such cases, it might be considered counterproductive to some training programs to formally identify performance shortcomings or problems, although evaluations are performed at some union-operated facilities at clients' requests.

Third, in contrast to the commercial air carrier and nuclear power industries, the nature of marine operations and differences in vessel configurations and maneuvering behavior results in considerable variability in the strategies used for successful operations. Thus, the approaches, practices, and techniques taught in a simulation course might not necessarily be those used aboard any particular ship. Simulator instructors may therefore be reluctant to say how an individual might perform aboard a given vessel.

Fourth, few, if any, simulators and simulations are currently validated across platforms. This potential inconsistency among platforms and simulations raises concern that the evaluation may not be an accurate reflection of the individual's true abilities and skills (see Chapter 7).

There is also concern that formal training performance evaluation records, including specific descriptions of course conduct and documentation of individual performance, could become evidence in accident investigations or disciplinary proceedings. The concern is that these records could be open to misinterpretation and that individuals might not wish to participate in training if records of their performance were retained. Given the nature of simulator-based training, and the fact that errors are routinely allowed to occur as a training tool, the concern over possible misinterpretation of training results by individuals not qualified to evaluate simulation performance should be considered when deciding whether formal records should be retained.

Employer Use of Simulation for Training and Evaluation

Hiring practices in the maritime industry, although not standard, are usually based on acceptance of a marine license as proof of basic competence. A newly employed deck officer is normally considered competent for initial employment and may be given significant responsibility the first day on the job based solely on possession of the required license. This immediate expectation of competence may exist even when the officer is serving aboard an unfamiliar vessel, in unfamiliar ports and waters, with an unfamiliar crew.

U.S. shipping and towing companies, however, usually do not promote deck officers to positions of greater responsibility based on a marine license alone. A period of sea service, often longer than that required by the international Standards for Training, Certification, and Watchkeeping (STCW) guidelines or USCG regulations, is normally required so a deck officer can acquire experience and skills needed for promotion and to give the employer and ship's senior officers time to observe and evaluate the deck officer's abilities.

Simulation could potentially be used as a tool for initial evaluation and indoctrination into a particular company's operating practices in routine and emergency situations, prior to actual engagement. This would eliminate the present concern in situations where an officer is "hired blind."

Use of simulation may also enable employers to shorten the period of on-the-job training required for promotion. Simulator training in company procedures and ship-specific operating practices, followed by an objective, performance-based evaluation of the skills acquired, appear to fill a void not previously addressed by traditional teaching and evaluating methods.

Some employers have already initiated standard programs that use simulation for training and evaluation. The Panama Canal Commission program, described in Box 5-1, includes formal pilot training using both onboard and simulator-based evaluation by senior pilots. Boxes 5-2 and 5-3 include comments on use of simulators for testing pilots and a summary of a simulator-based check-ride, respectively.

The Panama Canal pilot development is described in Appendix F. The program is considered unique and presents opportunities for measuring the effectiveness of formal pilot training. Potentially, the program could be studied to provide a resource for evaluating traditional versus simulator-based training.

> **BOX 5-1**
> **Use of Simulators for Performance Evaluation:**
> **The Panama Canal Commission**
>
> The Panama Canal Commission has a program for formal pilot evaluation based on a series of periodic shipboard check-rides by senior pilots. The Commission is presently testing the use of simulator-based evaluations to supplement the shipboard check-rides. The test program includes a series of "dual check-rides" wherein a pilot is checked one day aboard ship and tested a second day using the simulator. The tests are conducted by two different pilots, and the results of each check-ride are kept separate to ensure a blind test. The tests' results are being compared to determine whether there is a correlation between shipboard and simulator evaluations. Potentially, the program could be a resource for evaluating traditional versus simulator-based testing for licensing and other purposes.

LICENSING PERFORMANCE ASSESSMENT WITH SIMULATORS

Currently, simulator-based assessment in the United States is used directly in only two licensing assessments. The first is the radar observer certification mandated through ratification of the STCW guidelines, and the second is the recently approved master's level course (see Appendix F). This course was the result of an unsolicited proposal from the simulator operating company and is based on the successful completion of a training program.

The proposal that additional simulator-based programs for marine licensing assessment should be introduced was suggested to the committee during presentations. The suggestion was based on a belief that performance in such exercises would demonstrate not only the knowledge assessed in the written exam but also the application of that knowledge.

The *Licensing 2000 and Beyond* report (Anderson et al., 1993), discussed earlier, suggests that the use of designated examiners "for practical examination of individuals seeking lower level licenses where a demonstration of ability, in addition to or in lieu of written examinations, holds appeal. Similarly, a designated examiner used to sign off practical factors application of other license or certificates may prove desirable." Among the report's recommendations was "Place significant increased emphasis on approved courses, and other, more formalized methods of training and de-emphasize 'seatime,' un-verifiable for quality or quantity, as the principal guarantor of competency."

The committee, however, disagrees. It believes that before simulation can be effectively applied to the licensing process, there are a number of critical issues in both training performance evaluation and licensing assessment that should be

> **BOX 5-2**
> **Comments on Testing Pilots Using Simulators**
>
> I believe simulators can indeed be used for testing (based on our trial of simulator testing to date), but only for certain tasks, not every piloting task.
>
> We have recreated the same transit conditions on the simulator as in the canal. Our simulator experience showed that:
>
> - We need a lot of people involved—a master, helmsman, mate, and various voices on the radio—to simulate the signal stations, marine traffic control, other pilots, locksmasters, etc.
> - Depth perception is, of course, a problem, as is the field of view if it is less than 360 degrees.
> - We do not have such a large number of [simulated] ships, so there is a problem being able to have the same ship in the simulator test as the ship used for the check-ride.
> - As far as communication skills—understanding of canal operations, command presence, timing—the tests definitely show possibilities if viewed as separate tasks with clear, concise requirements.
>
> Captain S. Orlando Allard
Chief, Maritime Training Unit
Panama Canal Commission

addressed. As noted in *Licensing 2000*, "The use of simulators for testing purposes is controversial. . . . Further, wide-spread use of simulation for testing of more definitive subjective knowledge has yet to be fully demonstrated."

ISSUES IN SIMULATION EVALUATION OR ASSESSMENT

The committee has identified a number of issues and constraints that were viewed as impairing the broad, near-term application of simulation in the marine licensing assessment. Before a large-scale program is undertaken, these issues should be addressed.

The Need for a Systematic Approach

The recent decision of the USCG to add a simulator-based training evaluation to the master's licensing process was not systematic. Submission of an unsolicited proposal to conduct the course and its acceptance by the USCG suggest the need for a well-defined plan. To be adequately prepared for the quality

> **BOX 5-3**
> **Typical Summary of a Simulator-Based**
> **Check-Ride**
>
> The Panama Canal Maritime Training Simulator is a very effective tool to supplement the training of apprentices and limited pilots. A simulator, however, cannot replace the "hands-on" training the pilots in training (PITs) and pilot understudies (PUPs) receive while riding with another pilot on the canal. A simulator is useful for teaching shiphandling skills but cannot accurately reflect the true behavior of a given vessel in a given area of the Panama Canal. Therefore, the simulator should be used primarily as a teaching aid.
>
> **Advantages**
>
> With the simulation, PUPS, PITS, pilots, and shiphandlers are able to:
>
> - learn and practice proper canal communications,
> - reenact a transit in any environmental condition,
> - be exposed to the most common emergency situations, and
> - practice proper timing and approaches at locks.
>
> **Criticisms**
>
> The 10,000 dwt (deadweight ton) ship model does not handle in an identical manner to the standard 10,000 dwt vessels that transit the canal in the following ways:
>
> - The model has a greater turning radius.
> - The model requires greater RPMs to correct the vessel's heading.
> - The internship forces created by meeting ships had to be canceled. The forces established for this model were too strong.
>
> The artificial atmosphere in the simulator creates greater stress for the following reasons:
>
> - It is an unfamiliar work platform.
> - There are problems with depth perception. It is very difficult to judge.
> - The lack of motion. One cannot feel the vessel's true forward, backward, turning, and lateral motion on the simulator's fixed platform and visual projection.
>
> Jeffrey B. Robbins
> Pilot Training Officer
> Panama Canal Commission

control and oversight responsibilities of an expanded, simulator-based licensing program, the USCG must have a framework. Its core must be guided by formal simulator and simulation validation standards, applicable training course standards, and certification of instructors and assessors.

The USCG should also consider including in the framework provisions for follow-up of the currently approved master's course (and others as they are approved) to collect and analyze data, such as comparing success and failure rates in the new course to those of traditional testing methods.

From the information and conclusions in *Licensing 2000 and Beyond*, summarized in Chapter 2, it appears that the USCG's marine licensing infrastructure does not have the structure or staff to fully apply advanced testing technologies as an element of marine licensing or oversee the possible delegation of additional testing responsibilities to third or fourth parties.[4] Adopting a systematic approach would help ensure that the appropriate quality control and oversight infrastructures are in place and that all implementation issues discussed below are addressed.

The structure of the framework could make use of the instructional design concepts outlined in Chapter 3. Elements of instructional design that might be integrated include:

- characterizing populations for which marine licensing is required and specifying competency requirements based on specific task and subtask analyses;
- developing marine licensing goals and objectives;
- developing standard performance criteria to measure whether licensing goals and objectives are met;
- determining the knowledge, skills, and abilities (or ranges for each) required to meet standard performance criteria;
- determining requirements for practical experience needed to develop knowledge, skills, and abilities;
- developing examining and assessing methodologies, including matching assessment media to assessment objectives;
- identifying resource requirements and testing media and validating and correlating them with marine licensing objectives, including a detailed inventory of simulators available by type and an estimate of percent time potentially available for simulator-based licensing assessment;
- matching specific assessment techniques to licensing requirements;
- establishing assessor qualification, selection, training, and certification requirements necessary to ensure the quality of the marine licensing process relative to established objectives;

---

[4]In *Licensing 2000 and Beyond*, the term third party is someone who trains or teaches, and fourth party is someone, other that the USCG or a third-party trainer, who administers a test or makes subjective judgments about the competency of an individual applicant.

- reviewing and approving proposed programs and courses for satisfying marine licensing requirements, including course content and materials and qualifications and certification of assessors; and
- establishing a monitoring and program evaluation system for the marine licensing program itself to provide a basis for continuous improvement.

Developing a plan based on the instructional design process requires consideration of the issues outlined below, as well as the following two problems (discussed in Chapter 3), which are basic to the application of instructional design to the marine industry:

- An important element in applying instructional design to marine licensing is specification of competency requirements based on task and subtask analyses. As discussed earlier, although the maritime industry has conducted several task analyses over recent decades, the results of studies have not been widely accepted as accurate depictions of the skills, knowledge, or abilities of various license levels (third mate, second mate, chief mate, or master). The studies are now dated and do not specifically address behavioral aspects of individual job performance or the specific steps required. Many are general and unfocused regarding specific requirements of a particular fleet or class of vessel.
- The issue of range "ranges of acceptable professional performance" must be addressed. In contrast to related industries (aviation, nuclear power, etc.), there are no absolute performance specifications in the maritime industry. For example, the individual, professional judgment of a watch officer, pilot, or master determines when a rudder order will be initiated and at what magnitude. Another equally competent officer might choose a different course of action, and there are no standards for comparing the two as long as the results are equivalent and the vessel does not subsequently become involved in an incident or accident. Properly applying the instructional design process will require developing a methodology to address this problem.

The Development of Standards

The Need for Standard Evaluation and Assessment Scenarios

Among some people who met with the committee who are responsible for conducting simulator-based training programs there was a suggestion that the professional judgment or the subjective evaluation of a trained evaluator can probably be as effective in evaluating performance against stated criteria as are objective (criterion- or domain-referenced) measures. Their experience is that nearly all simulators, including manned models, uncover and display individual and team deficiencies regarding most aspects of navigation, watchkeeping, shiphandling, communications, and coordination. Failure to prepare appropriate plans, anticipate system failures, maintain situational awareness, and manage stress (i.e., many of the problems identified as contributing to marine casualties) will almost invariably show up in normal, everyday watchkeeping and shiphandling situations.

At present there are no standard simulator scenarios available for training performance evaluation or licensing assessment. Individual training establishments have developed their own scenarios, often in conjunction with individual shipping companies. For the *Master's Level Proficiency Course* (Appendix F), the STAR Center staff developed 10 generic scenarios for each of 4 exercises used in the simulator portion of the program. These scenarios, as well as the models on which they are based, were developed entirely in-house to standards determined at the STAR Center in consultation with the USCG.

For effective mariner performance evaluation or assessment on a simulator, industrywide standard evaluation and assessment scenarios are needed. They should be based on cross-platform and cross-student research. All parties of interest should be included in developing the standards for these scenarios (see Chapter 7 for a discussion of a mechanism for developing standards). They could be based on hypothetical (generic) information, real-world information, or a hybrid of both. Each approach has advantages and disadvantages with respect to use in training or marine licensing (see discussion of simulator types in Chapter 4).

The Need for Consistent Results from Simulators

In conducting any form of performance training evaluation or licensing assessment, it is important to distinguish between individual performance under real conditions and variations that are induced by the training environment or testing situation. The basis for making such a comparison is, for practical purposes, limited to expert opinion at present and in the foreseeable future.

In structuring any simulator-based licensing assessment program, it is very important to carefully define what levels of simulator validity are required or acceptable for different levels of licenses. Not all licenses require the high face or apparent validity possible with a full-mission ship-bridge simulator.

Before approving a simulator for formal training performance evaluations or licensing assessments, it is important, as part of the validation process, to have credible experts subjectively assess whether or to what degree a simulation consistently results in behavior that would be expected under identical or similar real-world conditions. Training-induced variations in individual behavior and performance would not necessarily disqualify the simulation as an evaluation media for a training program, but should be considered in applying evaluation methodologies. If a simulation is used for licensing assessment, it is crucial that variations in performance are recognized and accounted for during assessment. (Chapter 7 includes a more detailed discussion of simulator validity and standards-setting.)

Certification of the Evaluator or Assessor

If demands for formal simulator-based evaluation and assessment increase, many currently informal practices would need to be modified to a more systematic approach to ensure adequacy and consistency. Consideration should be given to formally separating the instructor and assessor roles, at least with respect to marine licensing practices. Such separation would help ensure the integrity of any training provided to meet licensing requirements.

For licensing assessment to be most effective, it must be conducted by impartial assessors or assessor teams who function separately from the simulator operation and who conduct the simulation exercises independently. (This mode of assessment is the opposite of what the USCG has done in the STAR Center's master's course.) Currently there is considerable competition among license-preparation schools regarding their ability to prepare mariners to pass license examinations. If licensing assessments were accomplished solely by individuals affiliated with the simulator facility, considerable care would be needed to avoid conflicts of interest. Separation of these functions would avoid the possibility of influencing the candidate's performance aboard the simulator.

The education, qualifications, and experience requirements for a simulator-based license assessor may be very different from those of a training program performance evaluator. Careful consideration should be given to defining the assessor's professional skills and competencies and how they can be measured, as well as judging his or her observational and assessment skills and competencies.

The assessor who is subjectively assessing the behavior and performance of a candidate in a license-granting situation must be capable of isolating, observing, and measuring that performance effectively and impartially. It may be necessary to develop specialized training or perhaps apprenticeships for assessors to ensure that they possess the skills required to make effective, impartial assessments.

In developing the licensing assessment framework, the USCG needs to ensure that they have, or can develop, the capability to qualify assessors. Currently, there is no formal certification of license assessors outside the USCG. The agency does have a system for instructor certification. As a part of the course-approval process, facilities are required to list instructors authorized to teach an approved course. If training (with or without formal evaluation of performance) were to be required as part of marine licensing, it is likely that company and union management would become more involved in observation and assessments.

Development and certification of qualified, impartial assessors will require time. It is important that the USCG consider a phasing-in process to allow time for this process to take place. Development of an adequate number of qualified assessors should be factored into the agency's licensing framework and phased approach to the introduction of simulator-based licensing.

Separating the Simulator and Student Performance

Related to the issues of simulation and simulator validity and to assessor qualifications is the concern of separation of simulator performance from student performance. In both training performance evaluation and marine licensing, it may be difficult to separate the performance of the simulator from that of the student or candidate. Full-mission simulators have a range of presentation capabilities (accuracy and fidelity), performance response (hydrodynamics and aerodynamics), physical layout (bridge hardware integration and design), and performance measurement (embedded). The older the system, the more likely that its embedded recording and reporting capabilities will be limited. The more recent installations offer wider choices of scale, frequency of plot, richness of detail, color graphics, and other features.

There have been no cross-platform studies as to the efficacy of specific platforms for training evaluation or licensing performance assessment. Furthermore, different platforms are limited to the ranges of performance they can simulate. These limitations include, for example, number of controllable other vessels, degree of fog or reduced visibility, hydrodynamic forces (e.g., channel and bank, squat, passing ship), and size of vessel team that may be accommodated.

In a training environment, a specific student may learn faster or more completely with one particular simulator-instructor combination than another. It may be that all simulators can meet the specific demands for performance evaluation and assessment if the criterion or domain to be measured is broad enough to factor platform constraints into the process. Some simulators may be more appropriate for certain types of training or assessment. Conversely, no simulator is best at all of these requirements.

Currently, student or license candidate performance is, at least in part, a function of simulator capability. Until cross-platform and cross-student (or candidate) studies have been performed, all simulator-based-training performance evaluation and licensing assessment must be recognized as being based on a particular simulator or simulator-evaluator/assessor combination.

Availability of Simulator Resources

The decision to require a specific level of validity for a specific level of license and to require use of standard simulation scenarios for specific license levels will have a major impact on commercially operated simulator facilities. The demand for testing could quickly exceed ship-bridge and other simulator resources. In developing a licensing framework that includes simulator-based training and assessment, the USCG should consider a phasing-in process to allow sufficient time for the marketplace to provide the resources needed to meet the potential demand.

Funding

Evaluations conducted in conjunction with training are a cost of doing business that can be accommodated in course structure. Evaluations and assessments for other purposes, if not conducted during training, could result in added costs. The required funding would depend on the testing platform and the manner in which the assessment or evaluation was conducted.

In *Licensing 2000 and Beyond* (Anderson et al., 1993), it is suggested that "in some areas it should be possible to shift expenditures [i.e., costs to the license or document applicant or his or her employer already incurred directly or indirectly in obtaining a license or document] from existing indirect costs into more constructively applied direct costs." This report suggests, for example, that these costs might be shifted "from present-day courses . . . to the cost of formal competency-directed training in an approved course. . . . Successful completion of the course could eliminate the U.S. Coast Guard exam."

Any USCG-mandated training or licensing assessment needs to include an analysis of the possible sources of funding to ensure that mariners have the ability to pay for the training, especially if the mandated training would effect license renewal and continued employment.

The recently approved master's course combines training and testing in a two-week period, and successful candidates receive their unlimited master's license. For this program, the American Maritime Officers Union has stated that all members seeking their master's license will be required to take this course, and the member's company will pay for the program.

Other Considerations

In addition to the technical issues discussed above, the following issues should to be addressed in the development of a framework for the USCG licensing program.

Implications of Combining Training with Formal Licensing Assessment

In addition to the concerns discussed above about combining the instructor/evaluator function with that of the assessor in marine licensing, consideration must be given to the development of adequate security measures at the operating and authorizing levels to ensure fairness, accuracy, and reliability.

Familiarization with the Simulator

No matter how high the fidelity and accuracy of a ship-bridge simulator, it is not a real ship. The individual whose competency is being evaluated or assessed needs to have some level of familiarity with the testing platform—whether it is a ship, ship-bridge simulator, manned model, or other form of simulator or testing device—so that the influence (positive or negative) of that platform on his or her performance is minimized. The degree and level of familiarization that should be conducted depends on whether the familiarization is used as training in advance of a performance evaluation or is in advance of an assessment required for marine licensing.

Familiarizing an individual with the simulator in conjunction with training is appropriate insofar as that which is being measured is performance outcome resulting from training. Using training to familiarize an individual with a simulator where the goal is competency determination for licensing is more problematic. There is a need for research to quantitatively determine whether extensive familiarization with a simulator artificially inflates individual performance. The effect of combining simulator training and simulator-based assessment within a two-week period is an aspect of the recently approved USCG *Master's Level Proficiency Course* that should be carefully reviewed.

An alternative to familiarization with a ship-bridge simulator through training is use of special indoctrination simulations designed for this purpose. This approach, however, would add cost without providing any additional return on the investment of time and resources, as might be achieved by using training as a vehicle for familiarization.

Role-Playing in Licensing Assessment

Conducting an effective performance evaluation or licensing assessment requires that all normal bridge positions are filled and are "played" in the same manner in which they would be aboard ship. Who should play these roles for evaluation and assessment is problematic and not easily resolved.

If colleagues play bridge team and pilot roles, especially if the colleagues are also to be tested, their performance in support of the individual being examined could mask weaknesses. Furthermore, colleagues may not represent the actual level of expertise that would be associated with each bridge team position in real life. Thus, the performance of colleagues is unlikely to exactly replicate what would occur at sea. (This consideration affects all role-players.) Personality conflicts might have the opposite effect by increasing the level of difficulty or interfering with the effective performance of the individual being examined.

Use of a simulator facility's employees for role playing is an alternative, but could also present problems. On one hand, consistency in evaluation or assessment could be created by using the same individuals as role-players, provided that their performance did not become so good as to implicitly enhance the candidate's performance. In competency determinations for licensing, however, the use of simulator facility employees, although perhaps economical, might result in the appearance of conflict of interest, because of the facility's interest in continued use of its resources for testing. In the USCG-approved master's course at the STAR Center, simulator facility employees were used as role-players while the candidate was being tested (see Appendix F).

Roles could be played by disinterested parties, in which case costs would likely increase, and there would be issues of quality control over the expertise of the role players. Another alternative would be for roles to be played by representatives of marine licensing authorities. Here again, there would be an overall increase in costs. The qualifications of the role-players might also be of concern.

FINDINGS

Summary of Findings

Understanding mariner training, competence, and proficiency, and the factors required in measuring human performance, either for real-world or simulator-based learning situations, is important for developing evaluation and assessment techniques. For the purposes of this report, the term evaluation is defined as being the output of a training program. Evaluations may be formal or informal, objective or subjective, or both. Assessment is defined as the output of a licensing process wherein the performance of the candidate is formally measured against a defined set of standard criteria. Assessments may also be objective or subjective. Evaluators work in training or performance evaluation programs, and assessors work in license-granting programs.

During the course of its work, the committee found that much of the evaluation done in connection with simulator-based training is informal. There are several reasons for this, including a reluctance to maintain formal training performance records because of the nature of simulator-based training, which can include the deliberate introduction of errors. A second reason is that many trainees already hold licenses and might be reluctant to attend training if they believed they were to be tested.

To date, simulation is used directly in licensing in only two instances—radar observer certification and in the recent USCG-approved *Master's Level Proficiency Course*. The USCG's acceptance of this latter course was in response to an unsolicited proposal. Before the agency undertakes more extensive use of simulation in marine licensing, program framework that includes consideration of the following issues should be developed:

- validation of the scenarios used in performance evaluation and licensing assessment;
- determination of the appropriate level of simulation for each license level;
- qualification and certification of the evaluators and assessors;
- availability of simulators; and
- source of funding for the cost of simulator use, especially in licensing assessment.

Two of these issues, evaluator and assessor qualifications and availability of simulators, could involve significant time. In the development of its licensing program framework, the USCG will need to include time to develop (or oversee development of) training courses and to train (or oversee training of) evaluators and assessors. The USCG will also need to factor in sufficient time to not only allow new simulator facilities to come on-line, but also to ensure that the simulators and simulations are properly validated.

One approach to developing the framework needed for the effective, widespread introduction of simulator-based licensing assessment would be to apply elements of the instructional design process discussed in the above section "The Need for a Systematic Approach." Systematic application of this process would address many of the issues raised by the committee.

Research Needs

In the course of its investigation of the uses of simulation for training performance evaluation and licensing assessment, the committee identified a number of areas where existing research and analysis did not provide sufficient information to extend its analysis.

The "Research Needs" section of Chapter 3 discusses the need to expand and update specific task and subtask analyses, including data on dimensions related to behavioral elements and specific steps needed to execute each subtask. In addition to using these data in developing training courses, they are important in developing the licensing assessment program. In use of simulators as part of the licensing program, the skills and abilities to be measured must be defined.

There are also few data regarding specific requirements of a particular fleet or class of vessel. Such information would be useful in applying instructional design to the simulator-based licensing assessment process.

Another concern of the committee is that in the marine industry there is little or no research that would assist in developing performance-based criteria to be used as measurements for determining whether simulator-based licensing objectives are being met.

Conducting licensing assessment using simulators at several different facilities would necessitate the development of industrywide standards for the assessment scenarios used. These standard scenarios should be based on cross-platform and cross-candidate research to ensure consistent, reproducible measures of performance.

There is also a need for cross-platform studies to determine the efficacy of specific platforms for training evaluation or licensing performance assessment. Individual platforms may be limited in the range of performance they can simulate, and they may be more applicable to certain levels of licenses. These limitations could include such factors as number of other controllable vessels, degree of fog or reduced visibility, hydrodynamic forces (e.g., channel and bank, squat, passing ship), and size of vessel team that may be accommodated. To judge which platform to use for which licensing assessment, it is necessary to understand the advantages and limitations of each type of platform, as well as the necessary levels of validity and fidelity for the different levels of license.

The committee could not find any research that would support or discredit the combining of simulator training and simulator-based assessment in a single course. The committee does, however, have some concerns about the potential problems (listed briefly above). These potential problems should be addressed, and research should be conducted to determine whether there are reasons to either combine or separate the two functions.

REFERENCES

Anderson, D.B., T.L. Rice, R.G. Ross, J.D. Pendergraft, C.D. Kakuska, D.F. Meyers, S.J. Szczepaniak, and P.A. Stutman. 1993. Licensing 2000 and Beyond. Washington, D.C.: Office of Marine Safety, Security, and Environmental Protection, U.S. Coast Guard.

NRC (National Research Council). 1985. Human Factors Aspects of Simulation. E.R. Jones, R.T. Hennessy, and S. Deutsch, eds. Working Group on Simulation, Committee on Human Factors, Commission on Behavioral and Social Sciences and Education. Washington, D.C.: National Academy Press.

6

Simulator-Based Training and Sea-Time Equivalency

INTERNATIONAL SEA-TIME REQUIREMENTS

Onboard experience has traditionally been required as a practical necessity for learning the skills of a merchant marine deck officer. The international Standards for Training, Certification, and Watchkeeping (STCW) guidelines standardized the sea-service requirements for deck officer licenses. It requires a minimum period of service in the deck department with prerequisite service in bridge watchkeeping duties before a prospective deck officer can be certified as officer in charge of a navigational watch.[1] There are additional onboard experience requirements for deck officers to upgrade to chief mate or master or renew a marine license or competency certificate. The International Maritime Organization's (IMO) role and the STCW guidelines are discussed in Appendix B.

The minimum period of onboard time, commonly referred to as *sea time*, is considered essential to ensure that mariners are exposed to actual operating conditions, over an adequate period of time, to prepare them for positions of increasing shipboard responsibility. Currently, the STCW guidelines require a minimum of three years of sea service for an original mate's license. Two of the required three years can be substituted by "special training" that marine licensing authorities are satisfied "is at least equivalent in value to the period of seagoing

---

[1]The international STCW guidelines distinguish among masters, chief mates, and officers in charge of a navigational watch. They do not distinguish between third and second mates because their at-sea watchkeeping duties are similar.

service it replaces. . ." (IMO, 1993). The STCW guidelines do not define the nature, form, or content of special training, an issue that is under review by the IMO.

The amount of time and structure of at-sea experience required to qualify for an original license varies widely, from the structured sea project required of U.S. Merchant Marine Academy cadets during their year of service aboard merchant vessels, to the three annual roughly 60-day training ship cruises of the state maritime academies, to the normally unstructured periods of at least 3 years at sea by unlicensed personnel qualifying for an initial license examination. Other than a requirement for documenting the duration of service, no national or international standards exist on the type of experience to be garnered during this period.

It is important to recognize that no baseline has been established as a frame of reference for determining adequacy of the current sea-time requirements for building the necessary mariner knowledge and skills. The minimum sea-time requirements have not been validated by either scientific research or documented by empirical evidence. Although analyzing the adequacy of the sea-time requirements is beyond the scope of this study, it is important to note that all discussions of sea-time equivalency are based on the assumption that existing requirements are adequate.

DEFINITION OF SEA-TIME EQUIVALENCY

The U.S. Coast Guard (USCG) and other certification authorities have begun to modify license requirements for onboard experience and have granted mariners *sea-time equivalency*—reductions in onboard service requirements—for completion of onshore training. The term *sea-time equivalency*, as used in marine licensing and this report, refers to a formal judgment concerning the relative value of simulator and other structured training as compared with actual hands-on experience gained aboard ship.

The concept behind sea-time equivalency is based on the perceived value of structured training, especially simulator-based training. During the past few years, the majority of mariners who have advanced to the level of master have done so through a combination of structured academic and onboard on-the-job training. The use of sea time equivalency represents a conclusion by the USCG that structured, simulator-based training is equivalent to certain on-the-job, shipboard experience.

The granting of credit for simulator training as a substitute for required sea time is referred to as *remission of sea time*. Recently, the USCG stated that it will encourage training and simulator use by expanding the practice of sea-time remission (USCG, 1993).

The STCW guidelines do not specifically define, though they do imply, that seagoing service is service aboard an actual ship. Remission policies are permitted, though not encouraged, under the guideline's present wording. As a matter of policy, the USCG has interpreted the existing language of the STCW guidelines as a sufficient basis for substituting time spent in ship-bridge simulator-based courses for sea-time remission ratios up to 6 to 1.

The application of a specific ratio (such as 6 to 1) implies that a judgment has been made that each day of the simulator-based training course(s) specified is equal to six days of on-the-job experience aboard ship. The USCG has extended the remission policy to include, within certain limits, portions of the minimum one-year sea time required by the STCW guidelines.[2]

During this study, the committee could find no strong support among mariners or employers for the desirability of remission of sea time. The committee also found that most groups were skeptical and ambivalent about the current policy. Pilot groups explicitly criticized the policy and recommended that it not be applied to pilots under any circumstances. This lack of support implies that, before additional measures are taken to increase sea-time remission, answers should be sought to the questions of whether, or to what degree, and under what terms and conditions simulation should be substituted for sea time in the professional development and qualification of mariners. The following discussions address these questions by considering unresolved issues, application of a systematic approach in the development of criteria for decisions to grant remission of sea time, and potentially useful applications of remission of sea time.

SEA-TIME EQUIVALENCY AND MARINER COMPETENCY

Chapters 2 through 4 discuss current simulator-based training and its applications to mariner skills improvement and professional development. There is, however, a basic, and quite important, distinction between the use of simulators as training tools and the use of simulator-based training for granting remission of sea time. Thus far, simulator-based training has been discussed in this report primarily in the context of a *supplement* to existing training practices. In the application of simulator-based training to the remission of required sea time, however, simulator-based training becomes a *substitute* for on-the-job skills and knowledge acquisition.

---

[2]The STCW guidelines are currently undergoing extensive revision, including major changes with respect to the use of simulation in the professional training of mariners. The draft revisions retain the three-year minimum sea-time requirement, the option for substituting training for two years of the requirement, and the minimum one-year sea-service requirement. The thrust of the proposed revisions is that the one year of sea service is an onboard requirement, although there may not be an absolute prohibition on the substitution of "equivalent" training. The final wording of the revisions, when adopted by the IMO, will need to be assessed to determine its effect on current and prospective remission policies and practices.

The primary goal of all mariner training is to improve *mariner competency* (including safety and efficiency of operations). That competency comes from both structured and unstructured training experiences. Current requirements for onboard service are based on the concept that certain professional skills development should take place through exposure to actual operating conditions and the resulting learning by experience and colleague-assisted training discussed in Chapter 1. When a decision is made to substitute simulator-based training for current practice, it is important to ensure that there will be no degradation of mariner competency and that safety will not be compromised. The ultimate test of every decision for remission of sea time, therefore, should be the test of the types of experience and skills being replaced by the simulator-based training and the possible effects of this substitution on mariner competency and the safety of the vessel, crew, and cargo.

BASIS FOR SEA-TIME EQUIVALENCY

Scientific Basis for Equivalency

The use of the word *equivalency* implies that simulator-based training replicates and equals on-the-job experience in terms of the development and reinforcement of the knowledge, skills, and abilities needed to perform effectively. A number of experiments conducted on the effectiveness of simulator-based training over the past 20 years concluded that the training is effective in its own right (Williams et al., 1980; Kayten et al., 1982; Multer et al., 1983; Miller et al., 1985; Reeve, 1987; Drown and Lowry, 1993). The committee could not, however, find any studies that directly addressed whether there is any equivalency between simulator training in selected skills and the skills learned ad hoc on board a ship.

A recently published study by MarineSafety International Rotterdam and TNO Human Factors Research Institute (DGSM, 1994) did attempt to determine what amount of simulator training is required to compensate for the first 30 days of cadets' sea time. The study compared two groups of students—one that had received only three weeks of simulator training and one that had completed their full sea time on board a ship. The study concluded that "the only ratio of simulator time to sea time that is based on observable data points, and not on extrapolations, is the ratio based on the 50 percent performance level." That ratio was 1 to 7.25, with a 95 percent confidence interval of 1 to 3.3 to 1 to 11.15. The application of this ratio yielded a conclusion that, for cadets, the 30 days of sea time could be replaced by 40 hours of simulator time (DGSM, 1994).

There is also some possibly supporting research in the commercial air carrier sector. Commercial air carrier research suggests that training in appropriate tasks can be accomplished more quickly, with effectiveness equivalent to on-the-job training, using well-founded, limited-task and full-mission simulator programs (Hays and Singer, 1989). It remains uncertain, however, that either the research in the commercial air carrier field, or the conclusions formed as a result of such research, are directly transferable to the maritime industry or other industries (Hays and Singer, 1989).

Anecdotal Basis for Equivalency

Acceptance of the concept that simulator-based training is equal to or better than certain on-the-job experience appears to be based almost exclusively on anecdotal information. In reviewing that information, it appears that all parties involved in simulator-based training share the commonly held conviction that it does improve job performance.

Professional mariners who have operated simulators as experienced trainers have expressed a strong conviction that "they know the method works." Students (the "end users") have provided feedback through course critiques, indicating that they believe that "simulator training is effective." Feedback from past students based on field experiences is, however, limited. As stated earlier, there is no information that effectively relates skills developed through simulator training to skills acquired on board ship.

ISSUES AFFECTING SEA-TIME EQUIVALENCY DECISIONS

The USCG has approached the issue of sea-time equivalency from a programmatic, not a technical, basis. USCG decisions to grant remission of sea time in ratios, such as 6 to 1, have been based on achievement of licensing objectives, a stated interest in encouraging training, and use of simulators or other program objectives. No technical basis has been developed for these ratios.

The USCG is using remission of sea time as an element of its licensing program. Its remission policies may have been ad hoc, based primarily on a perceived value of simulator-based training. It has authorized sea-time remission to assist the maritime academies in meeting the STCW sea-service guidelines and to encourage training.

Remission of sea time is, in fact, a complex problem that should be addressed much more systematically. Many of the issues discussed earlier in this report apply to considerations of sea-time remission. For example, issues raised in use of simulators for training, such as the need for a systematic approach to simulator training course development and the need for instructor qualification, should be considered in developing a rational approach to evaluating simulator-based training programs and their applicability to sea-time equivalency. Many of the issues related to performance evaluation with simulators and the application of simulation to licensing assessment should also be considered in decisions concerning sea-time equivalency.

Before any additional decisions are made in remission of sea time, these issues affecting possible implementation need to be addressed.

Mariner Tasks and On Board Equivalency

Earlier discussions of effective training and performance evaluation with simulators included an expression of the committee's concern that results of studies of the tasks and subtasks of mariners (in this case, deck officers and pilots) have not been used effectively and often do not provide information at the level of detail to adequately characterize the knowledge, skills, and abilities a mariner must develop and the behaviors and detailed steps involved in carrying out specific duties and responsibilities. To be able to define training objectives, then to measure through performance evaluation whether those objectives have been met, tasks and subtasks must be defined and matched to the ability and limitations of the simulator to train them.

It is important in the understanding of the tasks of mariners that all tasks are understood. Simulator-based training is not suitable for training every job skill needed on board ship. The random, sometimes high-stress situations that can develop during at-sea time are important for developing confidence, command presence, and interpersonal skills and for learning about the ship's business.

Sea time aboard ship provides a broad range of experiences that would be enhanced and complemented, but not replaced, by specific simulator training. As in other work environments, many tasks are often learned as part of routine duties without structured training. Deck officers and pilots learn these skills aboard ship at no additional expense to the mariner or operator.

In considering sea-time equivalency and the decision to identify specific simulator-based training as being equivalent to some period of on-the-job experience, this detailed understanding of the character of onboard tasks and subtasks being simulated is especially important. Without this understanding, it may not be possible to ensure that the substitution being permitted will result in achieving the level of mariner competency currently provided through onboard experience.

Training Course Standards

A related issue to sea-time equivalency is the quality of the specific simulator-based training courses that will substitute for sea time. If simulator-based training is accepted as equivalent to some amount of on-the-job training, it is necessary to be able to rely on that training as having accomplished specific training objectives as well as specific licensing objectives (see Chapter 5).

In the remission of sea time for cadets, the Maritime Academy Simulator Committee (MASC) has taken an active role developing quantitative data on which to base its recommendations that credit for 60 days of sea time be granted for successful completion of the 40-hour cadet watchstanding course. The proposed course structure would spread the 40 hours over 10 weeks to improve the effectiveness of the training. MASC is currently conducting a survey to compare shipboard and full-mission ship-bridge procedures to validate this proposed equivalency of 12 to 1.

Because the instructor must ensure that all course objectives are met, course instructor quality is also an important consideration. Certification or some other mechanism for ensuring instructor quality should be an element in any plan for determining sea-time equivalency. A part of that certification should be a requirement for instructor training. The three sample instructor training courses discussed in Chapter 3 are examples of efforts in this area.

Sea-Time Equivalency and License Levels

Because mariner knowledge and skill requirements vary by position and tasks, remission of sea time should be examined in the context of range of tasks and types of licenses. Granting remission of qualifying sea time for initial and upgraded licenses, for example, is entirely different from granting remission for recency of sea time for license renewal. An initial or upgraded license entitles the license holder to work in a new or different and, in most cases, more difficult position. A renewed license only permits a licensed mariner to continue working in the same position.

Consideration should be given, therefore, to remission of sea time in several different dimensions, including:

- original licenses issued to entry-level officers,
- original licenses issued to senior-level officers,
- upgraded licenses issued to senior-level officers,
- renewal of a currently held license,
- certificates of competency for specialized service, and
- credit for onboard trips to qualify for pilot licenses and extension of route.

Other Issues

Transfer and Skills Reinforcement

Before any major decisions are made concerning substitution of simulator-based training for sea-time requirements, the issues of transfer, retention, and reinforcement of learning from simulator-based training discussed in Chapter 3 need to be addressed. Transfer does not have to be complete for simulator-based training to be useful in improving skills. There should, however, be some mechanism to assess the effectiveness of the transfer and retention rate of skills learned. If transfer is incomplete, it may be necessary to ensure that the training is followed by some type of onboard reinforcement, such as an apprenticeship, to ensure continuing high levels of mariner competence.

Validation of Simulations and Simulators

In substituting simulator-based training for sea-time requirements, consistency in results can be crucial. Simulations are often modified in the regular course of doing business at many simulator facilities. Chapters 5 and 7 discuss the need for validation of simulators and simulations used in training and licensing assessment.

To ensure that training and licensing objectives are met, it is important to replicate simulator-based training results across platforms for like simulators. The training courses themselves, as well as the simulators and simulation scenarios used to substitute for sea time, need to be measured against established criteria to ensure that minimum standards are met. It is important that appropriate agencies recognize the need for standards and develop a mechanism to ensure that appropriate standards are set and enforced (see Chapter 7).

Cost and Availability of Simulators

Chapter 5 discusses the issues of availability of simulator resources and who would pay for simulator-based training that may be required in mariner licensing. Any mandated requirements for simulator-based training will result in a cost to the employer, union, candidate, or some other source. Part of planning for the expanded use of simulation in sea-time remission should be consideration of the potential impact of the cost of any mandated training.

The impact on the simulator facility "marketplace" should also be considered. Any proposed changes should be phased in to provide time for the marketplace to provide the resources needed to meet the increased demand that might result from substituting simulator-based training for sea-service requirements.

A SYSTEMATIC APPROACH TO DETERMINING SEA-TIME EQUIVALENCY

The decision to grant remission of sea time for specific simulator-based training is a decision to modify current licensing requirements. The committee could find no substantive technical basis for current USCG decisions that have been made in granting remission of sea time. It may be, however, that some decisions can be made despite the lack of a strong technical basis. It may be possible for the USCG to develop elements within its licensing program that grant remission of sea time in specific, well-defined areas, provided that the USCG program is based on a structured, systematic approach that includes a mechanism for ensuring that mariner competency and safety are not compromised.

The USCG cannot afford to continue to approach those decisions in an unstructured, ad hoc manner. Decisions to grant remission of sea time should be based on research that demonstrates the benefits and *only* after considering the issues outlined above. In determining whether to remit sea time, the agency should ensure that:

- Sea-time remitted clearly would not have provided as much benefit as the simulator-based training.
- There are some real, demonstratable benefits to the industry, public, or deck officer for the remission.
- It can be demonstrated that there is no degradation of other skills as a result of reducing sea time.
- Equivalency serves some purpose besides inducement to using simulator-based training.

It is important that the USCG develop a systematic approach based on research and an in-depth understanding of the type of experience replaced by the simulator-based training. To define the need for a policy of remitting sea time based on equivalency, the USCG should consider taking the following steps:

- identifying the license level and type (e.g., original licenses issued to entry-level officers) for which remission of sea time is being considered;
- outlining licensing objectives and competency requirements in terms of task and subtask descriptions, based on the license level and type defined;
- characterizing the tasks and subtasks most effectively developed through simulator-based training;
- developing criteria for simulator-based training based on licensing objectives;
- evaluating proposed simulator-based training programs against criteria;
- evaluating the full range of onboard experiences being replaced by simulator-based training;
- determining the relative importance of simulator-based skills in the mariner's overall work aboard ship;
- evaluating the degree of equivalency of each specific course and the specific transferred skills and determine how much time, if any, should be credited for the training;
- determining the skills learned aboard ship that are not covered by the simulator and the extent to which they deteriorate, if at all, as a result of sea-time remission policy; and
- establishing a mechanism for ongoing program monitoring through regular data collection and analysis to ensure maintenance of mariner competency and vessel, crew, and cargo safety.

Insofar as the value of the training course can be determined, it seems appropriate to vary the amount of sea time credited for a given course in accordance with its relative value in satisfying STCW guidelines and federal licensing requirements, as determined by qualified experts.

POSSIBLE EQUIVALENCY APPLICATIONS

As simulator-based training becomes more sophisticated and the objectives of the individual courses are more clearly delineated, the arguments become more persuasive that there is a basis for substituting simulator training for some presently required sea time. In fact, a number of applications of simulators could provide superior professional development. These possible applications need to be considered in any analysis of sea-time remission for simulator-based training.

Training for which Simulation is Well Suited

As discussed in Chapter 2, there are a number of attributes of simulator-based training that make it particularly effective in training certain skills. These attributes include:

- the ability to simulate adverse operating conditions safely,
- the ability to play back and repeat scenarios, and
- the ability to permit and use mistakes and accidents for teaching.

Because of competing demands of ships' scheduling and business, subjects such as emergency procedures, maneuvering in traffic, shiphandling, and bridge team and bridge resource management are sometimes given cursory treatment, at best, aboard many commercial ships. Simulator-based training could be superior to on-the-job training in these areas.

Encourage More Appropriate Training

Developing knowledge, skills, and abilities for prospective third mates is an area where simulator-based training may be considered an effective substitute for on-the-job training. Simulator training, for example, is substituted for a portion of the sea time required to prepare U.S. Merchant Marine Academy cadets for entry-level third mates' licenses. This training is considered preferable to traditional sea experience because the type of experience being replaced—service as unlicensed seaman in the deck department—is more effectively trained on the simulator. The ship-bridge simulator-based training develops watchkeeping skills and relates directly to the duties and responsibilities of a third mate's license (Hammell et al., 1980; Kayten et al., 1982). This substitution of sea-time requirement is applicable because the able-bodied seaman—especially aboard modern, more technically sophisticated ships—performs only a minimum of traditional seafaring tasks.

Simulator-Based Training to Meet Refresher Requirements

The STCW guidelines pertaining to frequency of training or refresher training are minimal, although there is an international trend toward more frequent training as a requirement for license upgrades and renewals. Current STCW guidelines are reflected in USCG regulations. The ship-bridge simulator can potentially be used to refresh knowledge, skills, and functional responsibilities in many areas for both license maintenance and recency purposes.

Simulator testing is much more performance-based than traditional, written examinations, particularly multiple-choice examinations used for license testing for the past two decades. Scenarios could be developed to demonstrate all desired skills and abilities—an advantage over shipboard checks (in those rare circumstances where they may be made) which can only demonstrate ability to deal with random situations. Furthermore, simulators could be used to train and check skills in emergency and hazardous situations that would never be deliberately created aboard ship.

Prior to recent developments in simulation, refresher training, which is now practical for many professional skills, could only be practiced aboard ship. Subjects that might be considered for refresher training using computer-based and manned-model simulation include:

- proficiency checks on nonshiphandling aspects of piloting for pilots;
- proficiency checks on shiphandling skills for masters and deck officers;
- new ratings (e.g., masters accepting their first-time appointments to very large crude carriers, pilots upgrading to bigger vessels or new ship types, masters appointed to vessels with nonstandard handling characteristics);
- bridge team management (masters and pilots, in particular, could be targeted for refresher training in human factors and bridge teamwork);
- bridge watchkeeping (for cadets and all watchkeeping personnel, to include rules of the road, navigation, voyage planning, and other skills inherent to watchkeeping);
- rules of the road (refresher training for masters, deck officers, and pilots); and
- shiphandling updating for pilots (manned models are preferred method).

Equivalency Standards for License Renewal

By using simulator-based examinations to test the renewing of mariner's skills in realistic situations, it may be possible to eliminate some or all recency requirements for officers renewing a marine license. Within the current system, it is assumed that a mariner's skills are current if he or she has a minimum period of service aboard ship, or in affiliated industries, in the years immediately preceding the five-year anniversary date of the mariner's license. As the U.S.-flag fleet shrinks, it is becoming increasingly difficult for ships' officers to maintain a shipboard career and obtain sufficient service to meet the recency requirements. Many mariners who want to maintain their licenses may be forced to let them expire.

In these instances, it might be reasonable to expect that a mariner who can demonstrate proficiency in realistic underway scenarios on a ship-bridge simulator has retained the basic proficiency level needed to renew an existing license. Recency may be academic, in this case, if continued proficiency is demonstrated in the simulator and a strong case can be made that a reasonably thorough demonstration of proficiency can be substituted for recent service.

In fact, substituting simulator training for recent sea time for license renewal may actually raise the standard of professional competence, while simultaneously reducing sea-time requirements, since credit is presently granted for service in rather loosely affiliated industries that often do not include any actual underway service or bridge watches aboard ship.

Providing Structured Training

At sea, on a commercial ship, a cadet or mariner gains experience randomly as individual situations develop—situations that are seldom selected to provide specific training. Although this real-world experience can be crucial in many training situations, the opposite may also be true for other training objectives.

In commercial at-sea training there is no structured instruction and often no structured monitoring or performance critique. On a simulator, the instructor and evaluator can teach and demonstrate preferred techniques and evaluate performance under a particular set of circumstances. Poor work habits can be corrected immediately.

On training vessels, operated by maritime academies, there is structured instruction and monitoring of cadet performance. The cadet is immersed in the total environment similar to a commercial ship. A training cycle on a training vessel would typically encompass a day of maintenance, a day of instruction, and a day of watch. Skills taught during class instruction on a training vessel and ashore at the academy are uniform. Modeling of the ship-bridge simulator after the training vessel and documents (voyage plans, check-off sheets, logs, standing orders, etc.) further enhances the training effectiveness.

Other Possible Applications of Sea-Time Remission

A modest level of sea-time remission might be granted for certain professional development courses taught using full-mission ship-bridge simulators. Given the lack of data on the effectiveness of the transfer of simulator-based training skills discussed in Chapter 3, any sea time granted in this instance would probably be based more on an intuitive relationship between onboard experience and simulator-based training than on quantifiable data.

FINDINGS

Summary of Findings

Strong maritime tradition has supported the position that onboard experience can be an effective way for mariners to develop the skills and knowledge necessary to advance to officer positions aboard merchant vessels. Recently, however, many mariners who achieve the position of master have advanced through a combination of structured and on-the-job training.

The international STCW guidelines have recommended minimum sea-service requirements for deck officers. The USCG, as the national license authority, has granted a remission of sea time in several areas—including license upgrades—for successful completion of specified simulator-based training. In granting these remissions, the agency has approached the issue of sea-time equivalency from a programmatic, not a technical basis. USCG decisions to grant remission of sea time in ratios, such as 6 to 1, have been based on decisions related to the achievement of licensing or other program objectives.

Recent work in the Netherlands concluded that in cadet training a ratio of approximately 1 to 7.25 would result in the substitution of 40 hours of simulator-based training for the first 30 days of sea time, at a 50 percent proficiency level. Beyond this study, the committee could find no other technical basis for these ratios.

It may be that for some specific applications, however, none is necessary. It may be possible for the USCG to develop an effective program for remission of sea-time by applying a systematic approach. Any decision to use simulator-based training for remission of sea-time requirements, however, needs to be treated similarly to use of simulation for licensing assessment. The same level of systematic analysis and assessment needs to be applied to both decisions.

A primary consideration of a sea-time equivalency program should be to ensure that mariner competency and marine safety are not compromised or degraded by substituting simulator-based training for sea service. The USCG should ensure that skills learned aboard ship that cannot be trained in the simulator do not deteriorate as a result of the remission of a sea-time system. The agency should demonstrate the benefits of sea-time remission through research and develop a system for granting such remission that includes consideration of all issues discussed and all steps outlined in this chapter, including mechanisms for:

- characterizating and comparing onboard tasks and subtasks to the objectives of the simulator-based training;
- developing training course criteria and performance standards,
- validating training simulators and simulation scenarios; and
- monitoring and improving programs continuously by ongoing collection and analysis of performance data and statistics.

There are a number of areas where simulator-based training may be more effective at achieving licensing objectives than on-the-job experience. These include:

- areas where onboard training may be limited by concerns over ship safety, such as maneuvering in traffic and shiphandling;
- cadet training;
- refresher requirements; and
- recency requirements in license renewal.

Until sufficient research is conducted to demonstrate the benefits of sea-time remission, it should be applied only to original third mates' licenses (for which simulator-based training has been demonstrated to be superior to the onboard experience) and to the renewal of licenses without grade increases (for which adding structured training to onboard experience would be a significant benefit).

Research Needs

Implementing a sea-time equivalency program will require a well-defined, ongoing monitoring program that will collect and analyze data to ensure that program objectives are met and mariner competency levels are maintained. Before there is any full-scale use of simulator-based training to meet certain refresher requirements, research should be conducted to determine the objectives of the current requirements and the type and length of training that could be substituted.

REFERENCES

DGSM (Director-General of Shipping and Maritime Affairs). 1994. Simulator Time and Its Sea-Time Equivalency. DGSM Project No. 634(70/3/017). MarineSafety International Rotterdam and TNO Human Factors Research Institute. Rijswijk, Netherlands: DGSM.

Drown, D.F., and I.J. Lowry. 1993. A categorization and evaluation system for computer-based ship operation training. Pp. 103–113 in MARSIM '93. International Conference on Maritime Simulation and Ship Maneuverability, St. Johns, Newfoundland, Canada, September 26–October 2.

Hammell, T.J., K.E. Williams, J.A. Grasso, and W. Evans. 1980. Simulators for Mariner Training and Licensing. Phase 1: The Role of Simulators in the Mariner Training and Licensing Process (2 volumes). Report Nos. CAORF 50-7810-01 and USCG-D-12-80. Kings Point, New York: Computer Aided Operations Research Facility, National Maritime Research Center.

Hays, R.T., and M.J. Singer. 1989. Simulation Fidelity in Training System Design: Bridging the Gap Between Reality and Training. New York: Springer-Verlag.

IMO (International Maritime Organization). 1993. STCW 1978: International Convention on Standards of Training, Certification, and Watchkeeping, 1978. London, England: IMO.

Kayten, P., W.M. Korsoh, W.C. Miller, E.J. Kaufman, K.E. Williams, and T.C. King, Jr. 1982. Assessment of Simulator-Based Training for the Enhancement of Cadet Watch Officer. Kings Point, New York: National Maritime Research Center.

Miller, W.C., C. Saxe, and A.D. D'Amico. 1985. A Preliminary Evaluation of Transfer of Simulator Training to the Real World. Report No. CAORF 50-8126-02. Kings Point, New York: Computer Aided Operations Research Facility, National Maritime Research Center.

Multer, J., A.D. D'Amico, K. Williams, and C. Saxe. 1983. Efficiency of Simulation in the Acquisition of Shiphandling Knowledge as a Function of Previous Experience. Report No. CAORF 52-8102-02. Kings Point, New York: Computer Aided Operations Research Facility, National Maritime Research Center.

Reeve, P.E. 1987. Mariner skills transfer by simulation. Pp. 35–38 in Proceedings of Problems of the Developing Maritime World, Sharjah, United Arab Emirates, January 26–28.

USCG (U.S. Coast Guard). 1993. Personal communication from USCG OPA90 staff to Committee on Ship-Bridge Simulation Training, National Research Council, July 23.

Williams, K., J. Goldberg, and A. D'Amico. 1980. Transfer of Training from Low to High Fidelity Simulators. Report No. CAORF 50-7919-02. Kings Point, New York: Computer Aided Operations Research Facility, National Maritime Research Center.

7

Simulation and Simulator Validity and Validation

The levels of trajectory accuracy and fidelity needed and delivered in the replication of ship maneuvering behavior for simulation training in computer-based and manned-model simulation are debated within the hydrodynamic modeling, marine simulation, and marine education and training communities. Ship-bridge simulators are used for all types of operational scenarios. Whether all of the appropriate vessel maneuverability cues are present in the simulation or correctly portrayed, whether the trajectory of the ship is actually correct, and the relative importance of accuracy in these areas are all important issues in the use of computer-based marine simulation (NRC, 1992).

Ship-bridge simulators are not only developed independently of the vessels that they simulate, they are routinely used to permit training in multiple hull forms and sizes. As a result, some simulator facilities use either a number of models to meet the specific application needs of training sponsors or adjust their models to simulate a different type or size vessel. If these adjustments are not correct, the resulting trajectory predictions will be inaccurate, regardless of the quality of the algorithms used to generate them or the apparent validity of the simulation. For these reasons, it is appropriate to validate each trajectory prediction model or perturbation to determine the capabilities and limitations of the product being delivered to the trainer, marine licensing authority, and licensing examiners and assessors.

The accuracy and fidelity of available ship-bridge simulators can vary significantly among facilities. These variations may result from differences in the mathematical models used to develop the scenarios or model modifications made by simulator facility staffs.

THE FIDELITY-ACCURACY RELATIONSHIP

As mentioned earlier, *fidelity* refers to the realism or degree of similarity between the training situation and the operational situation being simulated. The two basic measures of fidelity are *physical* and *functional* characteristics of the training situation (Hays and Singer, 1989). In the case of a manned model, the model contributes to both. In the case of computer-based simulation, the mathematical model contributes to functional characteristics. Fidelity is determined subjectively. The level of fidelity required is determined by the training objectives, which, in turn, are based on task needs and training analysis (Chapter 3; Hays and Singer, 1989). Determinant measures may be used to aid in assessing the level of fidelity in a given simulation. *Accuracy* is inherently a determinant measure of how close something is to being exact. The accuracy of a trajectory prediction model is determined by measuring variations of the predicted trajectory with the actual trajectory.

Correlating Realism and Accuracy

In many respects, fidelity is more difficult to address than accuracy because it involves a subjective assessment of how real the simulation is. Balancing accuracy of trajectory modeling with fidelity of motion in visual scenes, for example, is very challenging. It is possible to provide a believable simulation using a simple trajectory model that, with a few minor validating adjustments, can appear to be realistic to pilots and mariners in the specific harbor/ship situation. Yet, performing slightly different maneuvers than those used for validation can result in quite inaccurate trajectories. Indeed, all models have limited accuracy in various modes of which the trainer may be unaware. In general, this issue has not been addressed by simulation providers except to try to use the most accurate modeling approach economically available.

The accuracy of trajectory prediction models available to drive a simulation can be compared with the level of fidelity specified by the training analysis as necessary to achieve training objectives. The accuracy of trajectory prediction, for instance, is less important in courses where vessel maneuvering behavior is not an instructional objective than in courses where maneuvering is required to achieve the goal of certain learning situations or is the primary instructional objective.

Deliberate Departure from Realism

It is sometimes possible to enhance training effectiveness by departing from realism. As a general rule, in marine simulation, departures from realism are driven by limitations in training resources, rather than a conscious attempt to optimize training effectiveness. The most notable exception is the initial development of manned models, a development borne out of practical necessity to safely train the prospective masters of very large crude carriers in shiphandling. The scaling inherent in manned models is believed by many to enhance training effectiveness, although there are questions regarding the effect of scaling factors on individuals who do not have a well-established frame of reference in the operation of ships of the categories being simulated.

Because computer-based simulations rely primarily on software-based mathematical algorithms, there is considerable flexibility that could be used to deliberately depart from realism. In marine simulation, however, the opposite approach has been the rule. To build and improve confidence among mariners, training sponsors, and marine licensing authorities, there are strong pressures to use the highest level of realism possible. Nevertheless, it is possible to alter the mathematical trajectory prediction models to accentuate certain vessel maneuvering behavior, for example, as an instructional technique to assist a trainee in becoming aware of this behavior. As a rule, such an approach is problematic, because it appears that only a few ship-bridge simulation staffs have the level of sophistication in instructional design and hydrodynamic modeling to effectively stage and control deliberate departures from realism.

A major technical consideration in the application of simulators and simulations is the need for consistently reproducible results from simulation exercises. Currently, there are no standards for the development, operation, or modification of simulators or simulations. As their use is expanded from training to performance evaluation, licensing assessment, and substitution of training for sea time, consideration needs to be given to the establishment of industrywide criteria and standards.

PHYSICAL AND MATHEMATICAL SIMULATION MODELS

From the modeler's perspective, the simulation user must specify the accuracy needed for particular training or licensing objectives. The simulation modeler (for physical models or computer-based models) can then assess whether that accuracy can be provided. Pilots, for instance, need very accurate models to properly portray bank effects and other complex interactions, whereas a less-robust model may suffice to introduce very basic operational concepts and procedures to beginners. A detailed discussion of the hydrodynamic, physical models, mathematical models, and research needs in these areas is included in Appendix D.

Ship-Bridge Simulator Models

Current ship-bridge simulations are based on mathematical models derived by extrapolating hydrodynamic coefficients from towing-tank tests for a restricted

> **BOX 7-1**
> **Anchoring Evolutions:**
> **An Example of Needed Research**
>
> During anchoring maneuvers, the capability to determine changes in speed over the ground is very important. Anchoring evolutions have proven very difficult to simulate realistically in computer-based simulators. With respect to hydrodynamics, deceleration needs to be realistic under the influence of an astern bell. The retarding effect of dredging an anchor (that is, maneuvering about an anchor dragged along the bottom to retard the movement of the bow) needs to be correlated with propulsion and environmental conditions. The anchor needs to have the effect of holding the ship, dragging, and holding the bow as the ship swings into the wind. Understanding these effects and the ability to model them effectively is a weakness in the current state of practice of mathematical modeling and varies considerably among the various mathematical models.

set of hull shapes. It is generally assumed that the database is technically advanced enough to give accurate simulations of ship maneuvers in deep and unrestricted waters.

The simulation of ship dynamics for shallow-water effects, anchoring evolutions, and ship-to-ship interactions are less technically advanced. Simulations involving these factors are generally less accurate. The simulation of vessels being towed is also limited by an absence of systematic test data.

Numerous mathematical models are currently used to drive the various simulators. Hydrodynamic coefficients used in the models cannot be easily exchanged among most facilities. If one simulator facility has a need for a model of a ship it does not have, acquiring that model can be time-consuming and expensive. A standard method for exchanging models or modeling coefficients would facilitate sharing of important technical information. Development of model modules is also needed so that modules can be validated more effectively and an individual module can be replaced with one representing different vessel characteristics (e.g., propeller, rudder, hull trim, draft) without requiring elaborate modifications.

Inconsistencies in modeling capabilities can arise from several causes, including the lack of appropriate data and a lack of access to existing data. Additional research is needed to extend databases in areas such as ship maneuvering coefficients and anchoring evolutions (see Box 7-1). The International Maritime Organization (IMO) has set some criteria for collecting full-scale ship maneuverability data and has established a five-year collection effort. The IMO's intent is to revise requirements for the data based on knowledge gained in the process.

Applying computational methods to the determination of pertinent hydrodynamic parameters offers great promise as a solution to many current modeling problems. Practical developments and use in ship-bridge simulators cannot be implemented until existing research codes are applied to the specific computational tasks of maneuvering simulation. Two different levels of implementation might be possible.

At the first level, computational methods can be used to complement towing-tank testing in the determination of hydrodynamic coefficients. Examples where computational methods can play a useful role include ship-to-ship interactions in restricted channels and the use of more-elaborate computational methods that account for viscous effects to consider ship operations with small underkeel clearance in shallow water.

At the second level, computations of pertinent hydrodynamic effects could be performed with sufficient accuracy and speed to replace the traditional approach, which is based on curve fitting. This approach could be particularly useful for ship interactions and restricted water effects, where existing simulator models are severely limited. Based on three-dimensional panel methods, it is now possible to compute the relevant interaction forces and moments during the simulation, based on actual ship trajectories and channel topographies, thus avoiding uncertainties and limitations.

Full-scale real-ship experiments would advance the state of practice in modeling, particularly for shallow water and restricted waters with banks. Modeling of operations at slow speed and with reversing propeller situations also needs improvement. Vessels currently part of the U.S. Maritime Administration's Ready Reserve Fleet and some vessels in the Navy's Military Sealift Command fleet represent a possible source for data to validate and improve mathematical models.

A second source of information to improve simulation design could be the data bank of proprietary towing-tank information that exists at towing-tank facilities in the United States. Much of this data was developed for private clients and is currently unavailable for public use. If procedures could be developed to allow the public access to existing proprietary data *without disclosing the source*, simulation design might be significantly improved with a minimum investment.

Manned Models

Physical models provide a simple approach to training in the understanding and application of the hydrodynamics of ship motion in deep water and close-in operations, including docking and coming alongside, maneuvering in shallow water and near banks, and in ship-to-ship operations. The scaling of these ship dynamics, however, results in propulsion and rudder-action modeling inaccuracies. Also, because of the way human eyes spread when relating to the scale of the model, the trainee's stereoscopic vision is more acute on a manned model than on the ship. Currents, wind, and waves are also difficult to model and are usually done only to a limited extent, and then not at all facilities.

Simulation Software Issues

As with any large computer program, the software developed for ship simulation is subject to potential limitations. These include:

- errors or restrictions in the assumed hydrodynamic model,
- numerical errors due to the reduction of the model to computational form (such as time steps that are too large), and
- programming errors (bugs).

Over time, age and insufficient maintenance may effect a program's relative quality and relevance, unless special efforts are made to provide updated versions.

Computer programs developed for ship-bridge simulators require substantial investments of expertise and effort, and this investment is normally protected by licensing agreements. Further protection is often achieved by distributing the code in an executable form that cannot be modified or transferred to different computational environments. Public-domain software is preferable from a user's standpoint and offers the significant advantage that it can be shared within the simulation and hydrodynamic communities to hasten the exchange of ideas, correct errors, and improve hydrodynamic models. This concept of "open" software is particularly suitable if use of simulation becomes mandated (i.e., in licensing requirements). Access to "open" software would also facilitate validation of simulators and simulations.

CURRENT PRACTICE IN VALIDATION

As noted in Chapter 2, commercial air carrier simulators are evaluated and validated at a particular level, depending on their application, for a range of operating conditions. Generally, a commercial air carrier simulator represents a particular aircraft model with a cockpit arrangement specified by the operator. Air carrier simulators are re-evaluated every four months through the National Simulator Evaluation Program. Validations include both objective and subjective elements. The simulator's performance and handling qualities are evaluated according to engineering specification; pilot acceptance is determined through subjective validation.

Once the simulator has been validated, any changes in the performance characteristics necessitate revalidation. This approach is possible because simulators, which are generally developed concurrently with the airframe, are platform-specific and benefit from airframe development. (Marine simulators, which generally simulate generic ships and bridge equipment arrangement, are not ship-specific.)

Under present U.S. Coast Guard course-approval practices, formal validation of simulators and simulations is left to facility operators. There is no industrywide standard validation methodology. Typically, validation begins with the manufacturer of the simulator and its associated devices, hardware, and software. The manufacturer's performance evaluation is based on the manufacturer's internally generated criteria. Technical deviations (e.g., pixel size or color abnormalities or processing speeds) are adjusted to contract specifications. Because no specific vessel is usually being replicated and there are no international or fixed standards for bridge and engine room layouts, the location of navigation or communication hardware (e.g., radiotelephone, radar, automatic radar plotting aids, chart stand) are based on facility layout and management preference.

In general, simulator facilities qualitatively validate their simulators and simulations with respect to ship maneuverability and visual scenes and their integration and correlation. Evaluations of the accuracy and fidelity of the simulation image and vessel response characteristics are generally accomplished through subjective mariner testing. Several mariners with experience on ship types similar to those being simulated usually conduct their evaluation over a period of three to five days. During that time they are familiarized with the simulator and asked to evaluate the performance of the ship type under a variety of conditions. These conditions range from normal day and night operations, to abnormal, unusual, and emergency situations. The "test mariners" put the simulator through various maneuvers to determine the accuracy of the simulated maneuvering behavior.

Following each simulation run, manufacturer and facility personnel debrief the mariners, asking for descriptions of accuracies and inaccuracies in the simulated vessel's performance. The mariners may suggest changes in such factors as turn rate, response to rudder and engine commands, acceleration and deceleration, or visual scene relevance and accuracy. Facility operations personnel then make changes in the vessel or port database. This process continues on an iterative basis until the test mariners are satisfied, or until the operations personnel indicate that no more changes can be accommodated. This iterative model validation process represents the state of practice in the industry, but it is not endorsed by the committee.

FACILITY-GENERATED MODELS AND MODIFICATIONS

Development of Facility-Generated Models

Development of models by a facility's staff is not uncommon. They are often developed from modeling routines available in some ship-bridge software packages. The facility staff will develop its own models if they do not have one available in the software library acquired with the simulator or if the available model did not perform to the satisfaction of the trainees or the instructional staff.

Some individuals at facilities are self-taught modelers who create and then validate their own models without the benefit of an outside perspective to ensure overall reasonableness and accuracy of the simulated ship's maneuvering behavior. In addition, students are sometimes asked for suggestions, which can result in model modifications.

Field Adjustments to Mathematical Models

Typically, facility operators continue to modify simulators and simulations after the initial manufacturer and test mariner validations are complete. In some instances, the process of refining and modifying the model can continue indefinitely. There may be several reasons why a facility operator makes field adjustments in the mathematical models.

First, in using the model there may be anecdotal indications that the accuracy and fidelity of the maneuvering behavior are different from those needed to achieve training objectives. In some cases, staff or others associated with a simulator may "correct" real or perceived deficiencies in simulations of ship maneuvers by modifying the hydrodynamic coefficients. Because the mathematical models used to represent the maneuvers include many terms and coefficients, the danger exists that ad hoc modification of one or several coefficients to improve a particular simulated trajectory will actually degrade others. Such modifications are often based on the perceptions of a small number of users, and their judgments may not be accurate.

A second reason a facility operator may adjust a model is if the simulator manufacturer does not provide support for model corrections or if the support is not timely. In one example, a facility was dissatisfied with the performance of the library model supplied by the simulator manufacturer for a unique vessel category. The simulator manufacturer's arrangements for model refinement did not provide for a timely resolution. As a result, the facility generated its own model using a software routine that was a component of software provided with their simulator.

A large number of mariners were exposed to the facility-generated model and asked to evaluate its performance. The coefficients were then adjusted to generate performance they found acceptable. The facility reported that the resulting model produced more accurate vessel motions than the library model. Although mariner confidence was established in the simulation, the altered model was not reviewed by a hydrodynamicist to ensure its hydrodynamic accuracy and fidelity.

AN APPROACH TO SIMULATOR AND SIMULATION VALIDATION

At present there are no industrywide standards for simulators or simulations and no standard ship models; nor are there standard simulation scenarios for performance evaluation or licensing assessment. Without industrywide standards there can be no assurance that training or licensing assessment goals and objectives are being consistently met at different facilities and on different platforms.

Continually modifying the model—with the lack of consistency that results—could be problematic for students training on simulators because each successive class may be conducted with a slightly or even dramatically different simulation than an earlier class. The level of model accuracy required and the extent to which simulation accuracy needs to be validated depends on the proposed use of the simulation. The practice of continually modifying models and the resulting inconsistency will become a major concern if the simulations are used for formal performance evaluations, licensing assessments, or training that results in remission of sea time.

For validating models at the U.S. Merchant Marine Academy, the institution's simulator specifications contain the following:

> A wheelhouse poster containing general particulars and detailed information describing the maneuvering characteristics of each own ship modeled. This wheelhouse poster shall meet the requirements of IMO Res. A.601(15). In addition a maneuvering booklet shall be provided for each own ship modeled. This booklet should include comprehensive details of the ship's maneuvering characteristics and other relevant data. It shall meet the requirements of IMO Res. A.601(15).

Standards for Training Simulations

Generally, mariner instructors believe that cadet training, rules-of-the-road training, and bridge team and bridge resource management training do not require high levels of accuracy, only behavior that is qualitatively correct. To teach basic shiphandling in deep-water operating conditions, moderately accurate ship hydrodynamic models may be adequate.

High accuracy and fidelity in hydrodynamic, channel, and harbor models are important for shiphandling training for conditions of shallow water, confined water, and small underkeel clearances. High fidelity is generally required for marine pilots who perform at a much higher level of detail and precision in confined waterways than do ships' officers in general. Thus, in the context of pilotage training, and for specific ship and port operations, both the ship and channel-environmental models need to be as accurate as possible. Pilots will reject simulations they do not believe to be accurate.

Standards for Performance Evaluation or Licensing Assessment Simulation

For effective mariner performance evaluation or licensing assessment on a simulator, industrywide standards need to be established. To ensure consistency in scenarios used for evaluation and assessment, standards need to be developed based on carefully defined levels of simulation validity that is required or is acceptable for the appropriate level of evaluation or license being tested. Not all evaluations or licensing assessments require the high face or apparent validity possible with a full-mission ship-bridge simulator. Insofar as practical, the process should require the use of standardized mathematical ship models and operational scenarios and databases to ensure consistency.

To the degree that ship maneuverability affects individual performance during licensing assessment, standardized ship models should be used. The current lack of standardized mathematical models makes it difficult to evaluate mariner performance in ship maneuverability. Simulations used for licensing assessment and qualification of shiphandlers would benefit from being highly accurate with respect to trajectory prediction and overall realism.

A standard set of harbor operating conditions could be developed for specific ships that can be accurately validated against measurements of performance and used as a consistent basis for assessing mariner competency. Since accurate ship and channel-environmental models are required for some purposes, it would seem reasonable that such models be used for all purposes. Current mathematical models could be modular and validated in parts, so that accuracies could be established.

Once designed, the exercise scenarios need to be validated through cross-platform and cross-student research. For cross-platform validation to be possible, there need to be industrywide standards for the different levels and uses of the simulations. If a simulation is used for licensing assessment, it is crucial that any performance variations are recognized and accounted for during assessment.

If manned models are used for licensing assessment, validation will be needed to ensure that models produce faithful vessel behavior.

Objective and Subjective Validation

The approach to validation of marine simulators or simulations might follow the commercial air carrier industry's approach of including objective with subjective evaluation. Model validation requirements would be based on simulator-based training and licensing objectives.

Objective Validation

The elements of a simulation that require validation are the accuracy and fidelity of:

- image portrayal, including the content, quality, field and depth of view, and movement of the visual scene;
- the predicted ship trajectories based on hydrodynamic and aerodynamic modeling;

- own-ship (ship model) characteristics; and
- the operational scenarios used for evaluating or assessing.

These quantifiable factors can be measured and incorporated into an overall "realism rating" that can be compared among simulator facilities.

Ship maneuvering can be validated through a formal, objective process. Standardized models are selected and tested in towing tanks and the results compared to selected full-scale real-ship trials of the same ships to provide benchmark data for validation and testing of simulators.

Subjective Validation

Subjective validation of the simulator and simulation should extend beyond the "test mariner" program to involve the use of panels of credible experts to assess whether or to what degree a simulation consistently results in *behavior* that would be expected under identical or similar conditions in the real world. Such panels are especially important for any type of formal training evaluation or licensing assessment. Impartial validation panels need to be carefully composed of people from multiple disciplines, including, for example:

- a simulation instructor or assessor,
- a representative of the licensing authority,
- a subject-matter expert, and
- an expert to validate ship-model behavior.

An Approach to Validation of Ad Hoc Models and Modifications

There are strong reasons to discourage ad hoc creation or field adjustments of mathematical models using software within some simulators because field adjustments almost invariably introduce uncertainties and unanticipated inaccuracies.

The commercial air carrier industry requires that all modifications to a simulation be documented and that the simulator be re-evaluated after modifications are made. Within the marine industry, it may be possible to establish a simulation baseline through initial validation, then require documentation and revalidation as modifications are made. The program might include requirements such as:

- Creations or modifications of models must be undertaken by an interdisciplinary team of individuals qualified in both the operational and hydrodynamic elements of the simulation, or
- Field adjustments or changes in model behavior should be reviewed by a suitably qualified hydrodynamicist and an external mariner to validate hydrodynamic fidelity.

Procedures are needed to ensure that such changes are documented and that the original vendors of the data and appropriate authorities are notified.

Process for Developing Industrywide Standards

Standards can be issued by any organization whose authority is accepted in the subject area, either government or nongovernment, but they require a great deal of cooperation on all levels. "A regulation is a relatively short step from a standard, but an important one. Many times standards are converted to regulations with little or no change in text" (NRC, 1985). It is, therefore, very important that standards for simulators and simulations be carefully developed, including consideration of the views of all parties of interest.

There are a number of organizations actively interested in the development of simulator standards. Professional groups, such as the International Marine Simulator Forum and the International Maritime Lecturers Association, have been developing technical standards for simulators. The current revision of the international Standards for Training, Certification, and Watchkeeping guidelines is expected to contain guidelines for simulators, but is not expected to propose technical specifications in the near future (Drown and Mercer, 1995).

Within the United States, the process for developing standards is well developed. The American National Standards Institute (ANSI) is an organization whose principal function is to recognize and respond to needs for standards and to arrange for involvement by its standards-development members. ANSI can play a very effective role in serving as the focal point for communications between its members who develop standards and those in the private sector and government who need and use standards (NRC, 1985).

Other professional societies, such as the Society of Naval Architects and Marine Engineers and the American Society for Testing and Materials, might also have an interest in assisting in the development of industrywide standards. For simulator and simulation standards, there are numerous parties at interest, including facility operators, shipping companies, unions, pilot associations, port authorities, regulators, maritime academies, training schools, and the mariners themselves. To be effectively implemented, the simulator and simulation standards must be based on a consensus view.

It may also be of value for simulator manufacturers and facility operators to subject their activities to International Organization for Standardization 9000 quality assurance certification. Such certification would help ensure that minimum standards are being maintained by these organizations.

ISSUES AND FUTURE DEVELOPMENTS

Validity Issues

Because demonstrations might have high face validity but lack the rigor and thoroughness needed to determine the accuracy of the trajectory prediction capability, demonstrating the simulator or simulation is not a substitute for formal validation. Ultimately, in the use of simulation for training performance evaluation and licensing assessment, the employer or regulatory authority needs to consider which of the following factors bear on the simulation's validity:

- whether the trajectory prediction models were provided by a modeler or developed independently by the simulator facility;
- the manner and process by which the performance of the model or models were validated;
- the identity and credentials of the validators;
- whether the trajectory prediction model or models were adjusted since they were last validated;
- the accuracy that can be achieved with the accuracy prediction model;
- the quality of the bathymetry and hydrographic data that drive the waterway model used in the simulation (of special importance if the trainee will use the results of port- and waterway-specific training in that port or waterway) and how these data were obtained and validated; and
- the validity of the model's behavior.

Future Simulation Development

Vessel trajectory prediction modeling is a developed science that provides highly useful tools for building marine training simulators. From a technology perspective, the future is promising for improving the accuracy, flexibility, and extent of simulator-based training and licensing assessment applications. Theoretical and numerical methods are powerful and nearing practical application. Computational power, graphics, and multimedia capabilities and the proliferation of microcomputers enable the use of sophisticated mathematical trajectory models.

FINDINGS

Summary of Findings

The accuracy and fidelity of ship-bridge simulators can vary significantly from facility to facility. These differences derive from differences among original mathematical models used to develop the simulations and facility operator modifications to models after installation.

There are no industrywide simulator or simulation standards. Currently, simulations are initially validated by the manufacturer, then validated by the facility operator through use of subjective "test mariners." Often, facility operators continue to periodically modify simulations after the initial validation. This practice of continually modifying simulations can result in inconsistent training programs, as successive classes may be conducted with different simulations. These problems are of particular concern when a simulation is used for licensing or training related to remission of sea-time.

To address these concerns, simulators and simulations should be validated based on proposed use and required level of performance. All modifications should be documented and the simulation revalidated. The extent to which accuracy of a simulation needs to be validated should depend on its proposed use.

An approach to simulator and simulation validation might follow the commercial air carrier industry process, which relies on both objective and subjective validation. The objective validations could include development of a "realism rating" based on an assessment of factors such as image portrayal and predicted ship trajectories. The subjective validations could be conducted by an impartial panel of experts.

Many organizations and individuals have an interest in the development of these standards. There are a number of national and international organizations that could effectively work toward the consensus necessary to promulgate standards that would be acceptable to all parties.

Research Needs

There are a number of areas where additional research in hydrodynamic and mathematical modeling could be applied to improve the accuracy of simulators and simulations. On a process level, value could be gained from the development of a standard method for the exchange of models or modeling coefficients. Also, if models were developed in modules, validation of the simulations and simulation characteristics modifications would be facilitated, as would replacement of outdated modules. Finally, simulation models could be improved through greater public access to existing proprietary hydrodynamic and other towing-tank data.

Additional research is needed to extend databases in areas such as ship maneuvering coefficients and anchoring evolutions. Also, the application of computational methods for determining pertinent hydrodynamic parameters offers great promise as a solution to current problems in modeling and ship-bridge simulators. Ship modeling would be improved through development and implementation of research codes to be applied to specific computational tasks of maneuvering simulation.

Full-scale real-ship experiments would advance the state of practice in modeling, particularly for shallow and restricted waters with banks. Modeling for operations at slow speed and with reversing propeller situations also needs improvement.

REFERENCES

Drown, D.F., and R.M. Mercer. 1995. Applying marine simulation to improve mariner professional development. Pp. 597–608 in Proceedings of Ports '95, March 13–25, 1995. New York: American Society of Civil Engineers.

Hays, R.T., and M.J. Singer. 1989. Simulation Fidelity in Training System Design: Bridging the Gap Between Reality and Training. New York: Springer-Verlag.

NRC (National Research Council). 1985. Status of marine and maritime standards. Steering Committee on Engineering Standards for Marine Applications, Marine Board, Commission on Engineering and Technical Systems, Washington, D.C.

NRC (National Research Council). 1992. Shiphandling Simulation: Application to Waterway Design. W. Webster, ed., Committee on Shiphandling Simulation, Marine Board. Washington, D.C.: National Academy Press.

8

Conclusions and Recommendations

Simulation has been used to train mariners since the 1960s. Applied properly, simulators can go beyond the traditional test of knowledge to testing and assessing skills and abilities. Recent concerns about marine casualties and mariner competence and proficiency have lead the U.S. Coast Guard (USCG) to investigate the possibility of increased use of simulators in the programs under its jurisdiction.

The Committee on Ship-Bridge Simulation Training found that simulation can be an effective training tool, especially in bridge team management and bridge resource management, shiphandling, docking and undocking evolutions, bridge watchkeeping, rules of the road, and emergency procedures. Simulation offers the USCG an opportunity to determine whether mariners are competent in a much more comprehensive way.

Although there are not sufficient data to assess the full value or impact the use of simulators has had in altering or improving mariner performance, there is sufficient experience to warrant its continued and even expanded use. However, for the USCG to use simulation effectively for training and licensing, it is important that a stronger research base be developed and that the agency address issues of standardization and validation discussed in this report. The committee's conclusions and recommendations provide a technical framework for expanding the use of ship-bridge simulation for mariner training, evaluation, and licensing assessment.

USE OF SIMULATORS FOR TRAINING

Setting Standards for Simulator-Based Training Courses

The use of simulators as a tool for training mariners is increasingly accepted within the marine industry. In many cases, simulation has been added to existing training programs without substantial course redesign to ensure that the simulation contributes effectively to training course objectives. One result has been a lack of standardization in simulator-based courses.

The USCG plays a role in simulator-based mariner training, in part, through its incremental decisions to allow substitution of specific simulator-based training for sea-service requirements. To ensure that current mariner competency levels are maintained and improved, it is important that standards for simulator-based training courses used in licensing and remission of sea time be developed as soon as possible.

Recommendation 1: Marine simulation should be used in conjunction with other training methodologies during routine training, including cadet training at the maritime academies, for the development and qualification of professional mariner knowledge and skills.

Recommendation 2: The U.S. Coast Guard should oversee and guide the establishment of nationally applied standards for all simulator-based training courses within its jurisdiction. Standards development should include consultation with, and perhaps use of, outside expertise available in existing advisory committees, technical groups, forums, or special oversight boards. If the USCG relies on outside bodies, the process should be open and include interdisciplinary consultation with professional marine, trade, labor, and management organizations; federal advisory committees; professional marine pilot organizations; and marine educators, including state and federal maritime academies. Whatever process the USCG chooses to use should be acceptable from a regulatory standpoint.

Specific elements of the training standards should address:

- identifying training objectives;
- developing standard course syllabi;
- establishing instructor qualifications and certification (see Recommendation 3);
- codifying procedures for teaching watchkeeping, bridge team and bridge resource management, shiphandling, emergency response, and other fundamental tasks and skills;
- creating student evaluation and assessment methodologies; and
- structuring industrywide research and training program effectiveness measures.

Training courses in the USCG licensing and sea-time remission program should be required to meet the resultant training course standards. These standards are urgently needed and should be developed without delay. To guide the rapidly evolving state of marine simulator practice effectively, these standards should be developed within two years.

Recommendation 3: U.S. licensing authorities should require that instructors of simulator-based training courses used for formal licensing assessment, licensing renewal, and training for required certifications (i.e., liquid natural gas carrier watchstander, offshore oil port mooring masters) be professionally competent with respect to relevant nautical expertise, the licensing process, and training methods. The professional qualifications of the lead instructor should be at least the same as the highest qualification for which trainees are being trained or examined. Criteria and standards for instructor qualification should be developed and procedures set in place for certifying and periodically recertifying instructors who conduct training.

Use of Simulators to Promote Continuing Professional Development

Continuing professional development is an important element in acquiring and retaining knowledge and skills. Simulators can be extremely useful to deck officers and pilots for renewing and refreshing existing skills and acquiring new skills through exposure to new technologies and operational scenarios. Ship-bridge simulators can be effective for both deck officer and pilot training in bridge team management, bridge resource management, and some shiphandling, although current computer-based simulators are limited in ability to simulate ships' maneuvering trajectories in shallow and restricted waterways and ship-to-ship interactions—capabilities important to pilot shiphandling training.

Because of the increased use of U.S. waters by foreign-flag vessels, the USCG should extend its concern for mariner professional development beyond the United States. The International Maritime Organization (IMO) is currently revising the Standards for Training, Certification, and Watchkeeping guidelines, including development of guidance for use of simulators. As a representative to the IMO, the USCG has a role in international professional development.

Recommendation 4: Marine pilotage authorities and companies retaining pilot services should encourage marine pilots, docking masters, and mooring masters who have not participated in an accredited ship-bridge simulator or manned-model course to do so as an element of continuing professional development. Marine pilot organizations, including the American Pilots' Association and state commissions, boards, and associations, should, in cooperation with companies retaining pilot services, establish programs to implement this recommendation.

Recommendation 5: Use of simulators for professional development should be implemented on an international scale to enhance the professional development of all mariners who operate vessels entering U.S. waters and to reduce the potential for accidents. The USCG should advocate this strategy in its representation of marine safety interests to the IMO and other appropriate international bodies.

Special-Task Simulators

There are a number of applications where special-task or microcomputer simulators could be used effectively in mariner training. One of the primary limitations affecting the widespread use of microcomputers is the limited availability of desktop simulations and interactive courseware. Lack of readily available interactive courseware constrains the training community's ability to provide comprehensive training using modern training media.

Recommendation 6: The U.S. Department of Transportation should selectively sponsor development of interactive courseware with embedded simulations that would facilitate the understanding of information and concepts that are difficult or costly to convey by conventional means.

USE OF SIMULATORS IN THE USCG LICENSING PROGRAM

Need for a Plan to Use Simulators Effectively

There is a need for more-comprehensive, performance-based assessments of professional qualifications in the licensing of mariners. Continual reliance on structured, objective, multiple-choice examinations for determining that an individual has achieved a defined level of professional competency has four significant weaknesses:

- The process does not include a mechanism for assessing the candidate's ability to apply knowledge (i.e., no effective measurement of skills and abilities).
- The relevancy of the content of the multiple-choice examination to the real-life duties and responsibilities of a deck officer is limited.
- The present testing methodology does not test the officer's ability to prioritize tasks in response to dynamic, real-world situations, which ships' officers must do in the normal course of work aboard ship.
- Written examinations test skills sequentially, one at a time, so the officer's ability to perform several tasks simultaneously, as is routinely required aboard ship, cannot be determined.

Full-mission ship-bridge simulators offer the USCG an opportunity to make significant improvements in its mariner licensing program. Simulation can become an effective tool that goes beyond the traditional tests of knowledge to evaluate and assess mariner performance—competency, skills, and abilities. If simulation is used more in structured licensing programs, the need for a technical framework to incorporate it into licensing and other structured assessment programs grows more urgent.

Recommendation 7: The U.S. Coast Guard should develop a detailed plan to restructure its marine licensing program to incorporate simulation into the program and to use simulation as a basis of other structured assessments. Based on the results of the research detailed in Recommendation 16 (below), the program should include (1) identification of sections of marine licenses appropriate for assessment through simulators and (2) a methodology for regular assessment of program effectiveness in ensuring that mariner competency and safety levels are maintained. In developing the plan, the agency should consult with all parties of interest. Program objectives should include the following:

- improve the assessment of candidates for licenses,
- identify deficiencies in individual qualifications, and
- stimulate correction of deficiencies through supplemental preparation.

To establish the foundation for broad application of marine simulation in the licensing process, the USCG should:

- evaluate the examination subject matter based on a complete job-task analysis (see Recommendation 16);
- develop performance measures that can be effectively measured through simulation;
- develop a methodology for assessing license candidates using simulation, including an indication of the level of simulation needed for each level of license;
- develop standards for the validation of simulators and simulation scenarios (see Recommendations 14 and 15);
- develop professional standards and procedures for certifying and recertifying license assessors (see Recommendation 8);
- establish a mechanism to document and evaluate results; and
- use results of the evaluations to improve the program.

Recommendation 8: Licensing authorities should require that license assessors of simulator-based licensing examinations be professionally competent with respect to relevant nautical expertise, the licensing process, and assessment methods. Assessors should hold a marine license at least equal to the highest qualification for which the candidate is being tested or should be a recognized expert in a specialized skill being trained. Specific criteria and standards for assessor qualification should be developed, and procedures should be set in place for certifying and periodically recertifying assessors who conduct licensing assessments with simulators. The standards that result will require experience and credentials found only in the most seasoned members of the marine industry. It is not likely that people with that level of maritime experience will be found in the government. The USCG will probably need to look to professional groups and outside contractors for its license assessors.

Substitution of Simulator Training for Required Sea Service

The USCG program for granting sea-time remission has evolved ad hoc, not through a systematic technical analysis. An assessment of its sea-time remission policies should be included in the agency's plans for restructuring its licensing program. The previously noted need for standards for simulator-based training courses should be considered in developing a plan to allow substitution of simulator-based training for required sea time in cases where the committee finds such remission to be suitable.

The plan should include recognition that simulator-based training promotes the acquisition of skills and abilities in a safe environment using the repetition of scenarios to promote learning. In the case of third mate candidates, for example, structured simulator training can effectively replace some portion of the unlicensed sea-time requirement. For third mate candidates, the work on a modern ship undertaken during sea time often does not have much relevance to any of the work done by a third mate. For the cadet, that time might be better used to gain the requisite bridge expertise through a formalized program that combines structured simulator-based bridge team management and watchkeeping courses with an appropriate time at sea.

For license renewal without increase in grade, substituting simulator time for recent sea time can raise the standard of professional competence. For example, current USCG renewal practices grant credit for experience that is not directly relevant to the license. In some cases, that experience may have been gained in loosely affiliated industries and may not include any actual underway service or bridge watches. Substitution of simulator time would ensure that some minimum level of direct operational experience had been met at the time of renewal.

In addition, by its nature, reliance on ad hoc professional development during sea service does not ensure that an individual has received the necessary exposure or has learned from experience. Using an appropriately structured marine simulator-based training course that includes rules-of-the-road training, bridge resource management, passage planning, emergency procedures, and watchkeeping practices could provide an alternative method to measure whether continued competence has been achieved in these areas prior to license renewal.

The use of simulation has the potential to provide higher-quality preparation than the existing recency requirements. Any course intended to provide recency

CONCLUSIONS AND RECOMMENDATIONS   179

requirements needs to provide sufficient time, because learning occurs through repetition. A measurable understanding of retraining and refresher requirements necessary to maintain or restore professional competence to acceptable levels is needed.

The USCG currently grants remission of sea time for license upgrades for approved full-mission simulator-based training courses. Substitution of simulator-based training for upgrade to second mate, chief mate, or master, or for any type of pilot's license should not be considered until adequate research is available to demonstrate that simulator-based training can replace onboard experience and skills required for license advancement without degradation of mariner competency and the safety of the vessel.

Recommendation 9: The U.S. Coast Guard should grant remission of sea time for the third mate's license for graduates of an accredited, professional development program that includes bridge watchkeeping simulation. The ratio of simulator time to sea time should be determined on a course-by-course basis and should depend on the quality of the learning experience as it applies to prospective third mates, including the degree to which the learning transfers to actual operations. Research to establish a more formalized basis for these determinations should be implemented without delay (Recommendation 16).

Recommendation 10: The U.S. Coast Guard should establish standards for the use of marine simulation as an alternative to sea service for recency requirements for license renewal of deck officers and vessel operators. Remission of sea time should be granted for renewal purposes to individuals who have successfully completed an accredited and USCG-approved simulator-based training course designed for this purpose. The course should be of sufficient length and depth and include rules-of-the-road training, bridge team and bridge resource management, and passage planning. The ratio of simulator time to sea time should be determined on a course-by-course basis and should depend on the quality of the overall learning experience insofar as this learning transfers effectively to actual operations.

Use of Simulator-Based Training During License Renewal

Active mariners can benefit from structured training at each license level, and inactive mariners can refresh vital skills in a relatively short time through simulator-based training. In both cases, simulation can play an important role by ensuring that applicants demonstrate some level of baseline competence at the time of license renewal.

Current operating practices in many segments of the U.S. merchant marine do not routinely provide adequate opportunity for chief mates to acquire essential shiphandling and bridge team and bridge resource management experience and expertise, much less proficiency. Development of these skills is very important prior to receipt of a master's license and service as master. The use of manned-model and computer-based simulation can increase the chief mate's professional development by supplementing on-the-job experience and training with structured shiphandling and bridge resource management experience.

Recommendation 11: Deck officers and licensed operators of oceangoing and coastwise vessels who can demonstrate recent shipboard or related experience, but who have not completed an accredited simulator-based training course, should be encouraged to complete an accredited simulator-based bridge resource or bridge team management course before their license renewal. Those seeking to renew licenses who cannot demonstrate recent shipboard experience should be required to complete such a course before returning to sea service under that license.

Recommendation 12: Chief mates should be required to successfully complete an accredited hands-on shiphandling course prior to their first assignment as master. The license should be endorsed to certify that training has been successfully completed. Either a manned-model or a computer-based, accredited shiphandling simulation course should be established as the norm for this training.

Recommendation 13: Currently serving masters who have not completed an accredited shiphandling simulation course should be required to do so prior to their next license renewal. In addition, masters should be encouraged to attend an accredited shiphandling simulator-based training course periodically thereafter.

VALIDATION OF SIMULATIONS AND SIMULATORS

The effectiveness of ship-bridge simulation in training and evaluating students may be influenced by:

- the level and quality of courseware,
- the quality of simulation of the physical environment, and
- the ability of the simulator to effectively replicate vessel maneuvering behavior.

There are currently no standard criteria for accrediting simulators and simulator-based marine licensing programs. For simulation used in licensing and license-related training to be fully effective, it is important that there be industry-wide standards. It is currently common practice to make ad hoc adjustments to simulators and simulations. This practice leads to an absence of standardized ship models, simulators, and other simulation components. It is imperative that there be regular, critical analyses of all simulators and simulations intended to serve as examination or prerequisite training platforms for marine licenses to ensure that their degree of difficulty is equivalent.

Recommendation 14: The U.S. Coast Guard should enlist the assistance of standards-setting and other interested organizations and sponsor and support a structured process for validating and revalidating simulators, simulations, and assessment processes. In developing these standards, all parties of interest should be consulted. The validation process should include objective, subjective, and behavioral assessments. The objective validation should include quantifiable factors; the subjective and behavioral validations should be performed by an impartial, multidisciplinary team consisting, for example, of (1) the simulation instructor, (2) a representative of the licensing authority, (3) a subject-matter expert, and (4) an expert to validate ship-model behavior.

Insofar as practical, the process should require the use of standardized mathematical ship models and operational scenarios and databases to ensure consistency within the process and across facility platforms. Once a simulator, simulation, or assessment process has been accredited for use in licensing, recertification should be required for any significant adjustments that are made.

Recommendation 15: Staff at simulator facilities should have objective knowledge of the capabilities and limitations of the hydrodynamic models on which their simulations are based. Modifications of the coefficients to address real or perceived deficiencies should only be performed based on competent oversight by a multidisciplinary team. Procedures should be developed to ensure that such changes are documented and that notification is given to the original vendors of the data and to the cognizant authorities.

RESEARCH NEEDED TO IMPROVE MARINER TRAINING, LICENSING, AND PROFESSIONAL DEVELOPMENT

Development of a Quantifiable Basis for Assessing Simulator Effectiveness

Understanding the equivalency of simulation to real-life experience and on-the-job training is an important element in developing any comprehensive program for using simulators. Limited anecdotal evidence from the marine sector and from the commercial air carrier industry suggests that well-formed programs that combine instruction, simulation, team interaction, and debriefings can be used in a standard environment under quality-controlled conditions to develop theoretical and practical knowledge, as well as procedural and cognitive skills.

There are, however, insufficient data available that assess whether there has been effective application of the knowledge and skills derived from simulator-based training. There are also virtually no feedback mechanisms for assessing performance of past trainees, other than sporadic anecdotal reports.

The exact nature of the equivalency of simulation to real life has not been systematically investigated for several reasons:

- the existing job-task analyses are not adequate for this purpose,
- there has not been any systematic application of job-task analyses in either marine training or licensing for this purpose, and
- no systematic program currently exists to collect and analyze performance data for past participants in simulations.

Recommendation 16: The U.S. Coast Guard and the U.S. Maritime Administration, in consultation with maritime educators, the marine industry, and the piloting profession, should sponsor a cooperative research program to establish a quantifiable basis for measuring the effectiveness of simulator-based training. The research program should:

- document anecdotal reports of simulator-based training effectiveness to improve mariner performance and reduction of risk;
- acquire, update, and validate existing job-task analyses to be used in the establishment of experiential education and training requirements for professional development;
- analyze the degradation rate of specific knowledge and skills during actual service and while awaiting a billet;
- assess the transfer effectiveness of simulator-based training to actual operations;
- define the capabilities of full-mission, limited-task, and desktop simulators; and
- correlate the capabilities of each simulator type with job-task analyses to define the appropriate application of simulation to each job task in training and licensing.

With respect to pilot development, the research program should take advantage of existing pilot development programs.

Simulation of the Physical Environment

Ship control and navigation are visually supported tasks, especially in confined areas. Learning visual skills is an important process in developing proficiency in control and navigation. The relative brightness and size of lights are particularly important in night simulations when they may be the only useable visual cue for judging distance, speed, and location relative to shoals, shore objects, and other vessels. Distance judgments are also an important visual skill. In many simulators, the visual simulations are provided with a raster or projection technology. Raster systems have limited capabilities to represent bright lights, and projection systems result in distortion of distance perceptions as an observer moves around the simulated bridge.

In addition to visual systems, vibration, sound, and physical movement of the bridge in roll, heave, and pitch are ship operational characteristics that must be addressed by the bridge crew. The impact on training effectiveness of simulating these ship characteristics has not been verified.

Recommendation 17: The U.S. Department of Transportation (DOT) and the maritime industry should assess the impact on training effectiveness of apparent limitations in simulator visual systems. If these limitations have a negative impact on training effectiveness, DOT should encourage development of visual systems that overcome or minimize the negative aspects of current systems.

Recommendation 18: The U.S. Department of Transportation should undertake structured assessments of the need for simulation of vibration, sound, and physical movement. These assessments should include consideration of the possibly differential value of these various sources of information in different types of training scenarios.

Manned Models

In contrast to ship-bridge and radar simulators, manned models are a means of simulating ship motions and shiphandling in fast time. Manned models are an effective training device for illustrating and emphasizing the shiphandling principles. They are particularly effective in providing hands-on ship maneuvering in confined waters, including berthing, unberthing, and channel work. Manned models can simulate more-realistic representations of bank effects, shallow water, and ship-to-ship interactions than electronic, computer-driven ship-bridge simulators.

There are currently three operating manned-model training facilities worldwide, one each in England, France, and Poland. The fourth, a U.S. Navy facility in the United States at Little Creek, Virginia, was recently closed. During its operating life, the U.S. Navy's manned-model facility provided opportunities for hands-on shiphandling training and instructor experience for some merchant mariners in conjunction with U.S. Naval Reserve service. Training at the facility also exposed some USCG junior officers to the maneuverability of tankers and cargo ships. Recent closure of this facility represents the loss to the United States of a unique training resource. Although the contribution of this facility to operational safety of commercial vessels was a side benefit, it nevertheless filled a gap in U.S.-based training resources for the development of merchant mariners.

Recommendation 19: Because there are no manned-model training facilities in the United States, and because of the usefulness of these models in familiarizing pilots and others with important aspects of shiphandling, DOT should study the feasibility of establishing or re-establishing a manned-model shiphandling training facility in the United States, to be operated on a user-fee basis.

Vessel Maneuvering Behavior

The ability of a simulator to closely replicate a ship's maneuvering trajectory is a strong measure of the usefulness and value of the simulator for training and licensing. At present, computer-based simulation of a ship's maneuvering trajectory is well developed in normal deep-water, open-ocean cases. In cases involving shallow or restricted waterways, ship-to-ship interactions, and extreme maneuvers, fidelity may be significantly reduced.

Recommendation 20: The American Towing Tank Conference (ATTC) and International Towing Tank Conference (ITTC) should be advised of the needs to extend the database of ship maneuvering coefficients. ATTC and ITTC should be encouraged to investigate possible development of procedures that would allow the exploitation of existing proprietary data without source disclosure. Where data are not available from these sources, funds should be allocated to perform new tests, especially in very shallow water and in close proximity to channel boundaries or other vessels.

Improvement in Mathematical Models

The current practice for developing mathematical models for simulators is based on extrapolation of hydrodynamic coefficients from towing-tank tests for a restricted set of hull shapes. This practice may degrade the validity of the information when applied to conditions of loading and trim or to ships that differ from the model tests. In addition, the simulation of towed vessels is severely limited by the absence of systematic test data. Conducting full-scale real-ship experiments would significantly advance the state of practice in the development of mathematical models.

In general, computational methods for determining the pertinent hydrodynamic parameters based on theories offer the possibility of more general and accurate simulations, particularly for ship operations in restricted waters and ship-to-ship interactions.

Recommendation 21: The U.S. Department of Transportation should develop standards for the simulation of ship maneuvering. The fidelity of the models should be validated through a structured, objective process. Standard models should be selected and tested in towing tanks and the results compared to selected full-scale real-ship trials of the same ships to provide benchmark data for validation and testing of simulators.

Recommendation 22: The U.S. Department of Transportation should initiate research to integrate computational hydrodynamics analysis with simulators in real time.

FUNDING SIMULATOR-BASED TRAINING AND LICENSING

Specialized training on manned-model and computer-based simulators is not affordable to most individual mariners. Improvements in mariner competence and professional development possible through simulator-based training have been discussed throughout this report. Professional development is a shared responsibility among the mariner, shipping companies, unions, port authorities and facility operators, and others. Each of these groups, as well as the general public, benefits from improved mariner competency and safety.

It is important, therefore, that any decisions to mandate training for licensing or renewal include full consideration of options and mechanisms available to ensure that the training is affordable and available to all effected mariners.

Recommendation 23: The U.S. Coast Guard and the U.S. Maritime Administration should assess the options for funding simulator-based training and licensing. The assessment should include the following options:

- allocation of some customs fees collected on marine cargos for maritime professional development,
- federal surcharges on port fees and pilotage fees,
- federal port fees,
- employer contributions,
- collective bargaining agreements,
- nonprofit foundations and other organizations, and
- individual cost sharing.

APPENDICES

APPENDIX
A

Biographical Sketches of Committee Members

WILLIAM A. CREELMAN, *Chair*, U.S. Maritime Administration (retired), received his B.S. in nautical science from the U.S. Merchant Marine Academy. He held a Coast Guard First Class Pilotage License for various East Coast waters. He was appointed as deputy maritime administrator in 1988, having previously served as deputy maritime administrator for inland waterways and the Great Lakes. Previously, he was president of National Marine Service (NMS), a towing industry company carrying bulk liquids on the inland waterways, completing 30 years of service in 1985. Afloat service with NMS included pilotage of coastal tankers on the East Coast. He served twice as chairman of the board of directors of the American Waterways Operators (AWO), served as chairman of various AWO committees and task forces, and also served as vice chairman of the Coast Guard's Towing Safety Advisory Committee. Mr. Creelman was a member of the Marine Board of the National Research Council.

PETER BARBER is senior lecturer, Maritime Simulation Section, Southampton Institute of Higher Education, Warsash, U.K., where he specializes in ship-handling training using manned models, bridge team management training on full-mission ship-bridge simulators, radar, vessel traffic system training, and port development contracts. He received maritime training from the College of Maritime Studies, earning his Master Mariner Certificate in 1976. Mr. Barber has been an instructor and lecturer for a full range of computer-based and physical scale manned-model training simulations. He participated in an assessment of radar simulation use as part of COST 301, the European Economic Community's broad examination of vessel traffic services. His consultancies included port development simulations for ports in Australia, the United Kingdom, and Hong Kong. He returns to sea regularly in an advisory capacity with various shipping companies and pilotage associations to update his knowledge and skills. Prior to joining the Warsash faculty, he served as chief and first officer with Townsend Thoresen Ferries. Earlier, he was first officer with Cunard Steamship Company, serving as training officer aboard the *Queen Elizabeth 2*. Prior to this position, he served in all ranks to chief officer on a broad range of ships in the British Merchant Navy. Mr. Barber's publications include use of shiphandling simulations for training and waterway design and simulation training for operators of vessel traffic services. He is a member of the Nautical Institute.

ANITA D'AMICO BEADON is assistant director, research and development, Grumman Data Systems, where she manages projects related to advance technology hardware and software. She earned her B.A. in psychology from the University of Pennsylvania and her M.A. and Ph.D. in psychology from Adelphi University. She has also served as senior human factors engineer at Grumman Aircraft Systems, where she developed functional requirements for rapid display prototyping software used to design and test new display concepts for cockpits and space vehicles. She is also assistant research professor, Adelphi University, where she teaches courses on advanced research design and execution. Previously, as an employee of Eclectech Associates, a division of Ship Analytics, she was assistant program manager and research director at the Computer Aided Operations Research Facility (CAORF) of the U.S. Maritime Academy. While at CAORF, she conducted and supervised research in simulation training, simulation fidelity, layout of bridge instrumentation, design of navigational aids, design of ports and harbors including the Panama Canal simulation study, watchstanding fatigue, and work scheduling. Her consultancies have included research on watchstanding officer workload for the SUSAN ship simulation facility in Hamburg, an ongoing study of drug-induced fatigue on worker productivity, and numerous research in experimental methods, statistics, and programming. She has had various other faculty appointments with Adelphi University, the State University of New York at Farmingdale, and Molloy College. Previously she was a research assistant at Adelphi University, the Temple University Medical School, and the University of Pennsylvania. Dr. Beadon has authored and coauthored numerous technical reports and published extensively. She is a member of the Human Factors Society, SAE Aerospace, Association for Computing Machinery-Special Interest Group on Computer-Human Interaction, and the Society for Information Display.

PETER H. CRESSY, rear admiral, U.S. Navy (retired), is the chancellor of the University of Massachusetts, Dartmouth, and the former president of Massachusetts Maritime Academy. He earned his B.A. from Yale University, his M.S. in international affairs from George Washington University, his M.S. in systems analysis from the Naval War College, his M.B.A. from the University of

Rhode Island, and his Ed.D. in organizational development from the University of San Francisco. He was awarded an honorary Ph.D. in organizational psychology by the Pacific Graduate School of Psychology. He has extensive experience in naval training and development and manpower utilization, including a Navy-wide appraisal of training requirements. His naval service included positions as commander, Fleet Air Mediterranean, Naples, Italy; director, Total Force Training and Education Policy Division; and director, Naval Aviation Training and Manpower Division, Washington, D.C.; commanding officer, Anti-Submarine Warfare Training Group Atlantic; commanding officer, Patrol Wing Five; and various legislative and U.S. State Department assignments. He has also taught courses for the Navy and at nine colleges and universities and developed curriculum publications. He sits on the academic board of the Pacific Graduate School of Psychology and chairs several marine boards and seaport councils in Massachusetts. He completed naval service as rear admiral.

DOWARD G. DOUWSMA is a principal in the Grafton Group, Gainesville, Georgia, where he specializes in organizational strategy and development. He is also associate professor of business administration at Brenau University where he teaches graduate and undergraduate courses in corporate strategy and marketing. He earned his B.A. in political science and economics at North Central College, his M.B.A. in strategic systems management at Baldwin Wallace College, and his Ph.D. in organization and management at The Union Institute. He has consulted in simulation-based learning for the Maritime Training and Research Center, Toledo, Ohio; the Maritime Institute of Technology and Graduate Studies, Linthicum Heights, Maryland; and nonmaritime firms including General Motors Corp., Midland Ross Corp., and Pickands Mather and Co. He developed and presented vessel/bridge resource management courses for U.S.- and foreign-flag fleets as well as the U.S. Coast Guard. He has conducted funded research using simulation for government agencies. He served as an intelligence officer in the U.S. Navy during the Vietnam conflict. Dr. Douwsma has authored and coauthored a number of articles and has made presentations of national and international focus on the use of ship-handling simulation for training and waterway design, including the 1991 Transportation Research Board annual meeting. His memberships include the Association for Management of Organization Design, the Society for Computer Simulation, and the Society of Naval Architects and Marine Engineers (affiliate member).

PHYLLIS J. KAYTEN is scientific and technical advisor for human factors at the Federal Aviation Administration (FAA). She is currently assigned to the NASA Ames Research Center, Palo Alto, California, where she functions as liaison between the FAA and NASA on human factors issues and research. She earned her B.A. in psychology from Brandeis University and M.A. and Ph.D. in developmental psychology from the State University of New York at Stony Brook. She provides highly specialized scientific and technical advice and guidance in aviation human factors development programs, including training system design, control and display design for cockpit and air traffic control systems, personnel selection, workload, impact of automation and new technology on human performance, and selection criteria measurement of individual and group performance. Previously, Dr. Kayten was special assistant to John Lauber, member, National Transportation Safety Board (NTSB), for whom she provided technical support on human factors issues in transportation accidents. She earlier served as human performance investigator on the NTSB staff for on-scene investigation of railroad, highway, marine, and aviation accidents, including examination of training and operation factors on human performance. Prior to these positions, Dr. Kayten was research analyst for human-computer interface and simulator training validation at Ship Analytics. She designed and conducted research at the Computer Aided Operations Research Facility that provided the basis for Coast Guard accreditation of simulator training of Merchant Marine Academy cadets as equivalent to a portion of the required sea time for deck officer licensure. Dr. Kayten has authored and coauthored a number of technical reports and papers.

GAVAN LINTERN is associate professor and associate head, Aviation Research Laboratory, Institute of Aviation, University of Illinois, where he instructs on human factors, engineering psychology, and cognitive science. He received his B.A. in psychology and mathematics and M.S. in psychology and mathematical statistics from the University of Melbourne, Australia, and his Ph.D. in engineering psychology and computer science from the University of Illinois. He specializes in flight training research. His research includes the impact of simulator design features, such as visual scene detail, and training effectiveness and the value of special instruction, such as part-task training, to flight training. He has been a principal investigator for Canyon Research Group with the U.S. Navy's Visual Technology Research Simulator, Naval Training Equipment Center, and has worked as an experimental psychologist in human factors with the Aeronautical Research Laboratory in Australia. He also served 11 years with the Australian Army. Professor Lintern has published extensively. He is on the editorial board of *Human Factors* and is a member of the Human Factors Society and the International Society for Ecological Psychology.

DANIEL H. MacELREVEY is a senior pilot with the Panama Canal Commission. He is also a principal in Offshore Services Company, a consulting firm specializing in shiphandling, training, marine technical writing, and services, as an expert advisor for admiralty law cases involving shiphandling and vessel operations. He earned his B.S. (cum laude) in nautical science from the U.S. Merchant Marine Academy. He served progressively from third mate to chief mate aboard vessels of the Moore-McCormack Lines and qualified for a master's license (oceans, any tonnage) in 1970. He qualified as Panama Canal pilot (all waters, any tonnage) in 1972. He has participated in a number of computer-based simulations. He subsequently served ashore and afloat with El Paso Marine Company for four years, including command of one of that company's LNG ships. Earlier, Captain MacElrevey authored *Shiphandling for the Mariner*, a leading book on the art of shiphandling presently in its third edition. He also authored numerous manuals and guides on shiphandling, marine operations, and training and watchkeeping for El Paso Marine Company and the Panama Canal Commission, along with various other marine publications. Captain MacElrevey served as an officer in the U.S. Naval Reserve. Memberships include the Council of American Master Mariners; the Pilot Division of the International Organization of Masters, Mates, and Pilots; and the Panama Canal Pilot's Association.

EDMOND L. MANDIN, American President Lines (APL) (retired), received his B.S. in nautical science from the U.S. Merchant Marine Academy. He provides marine operations consulting services to APL and other clients. Captain Mandin joined APL in 1948, and was assigned as master in 1961. He commanded 17 APL ships on worldwide trade routes. Appointed marine superintendent in 1987, he initiated and implemented companywide deck department personnel standards, produced and instituted training criteria, developed bridge team management and emergency response team underway training, and contributed to the development of a pilotage training curriculum. He also originated and implemented a system of vessel handling in heavy weather through applied meteorology, wave motion, and ship response techniques. His memberships include the International Organization of Masters, Mates, and Pilots and Council of American Master Mariners, of which he was a founding member of a local chapter.

ROBERT J. MEURN is currently full professor, Department of Marine Transportation, U.S. Merchant Marine Academy (USMMA), Kings Point, New York. Master mariner and captain, U.S. Naval Reserve (retired), Professor Meurn received his B.S. in nautical science from the USMMA and his M.A. in higher education from the George Washington University. He taught at the Texas Maritime Academy, was commandant of cadets and executive officer of the TS *Texas Clipper*, and was selected by the USMMA as teacher of the year in 1978. In 1981 he initiated the first watchstanding course in the United States for cadets, using the Computer Aided Operations Research Facility simulator. In 1983 he was honored again as teacher of the year at USMMA, where he served as head, Nautical Science Division. He coauthored the second edition of *Marine Cargo Operations* in 1985 and authored *Watchstanding Guide for the Merchant Officer* in 1990 and *Survival Guide for the Mariner* in 1993. In 1995 he was the recipient of the U.S. Department of Transportation's Bronze Medal award for his contributions to safety in marine transportation. Professor Meurn has sailed with U.S. Lines, Farrell Lines, American Export Lines, Moore-McCormack Lines, Grace Lines, and the Military Sealift Command as relief master and chief mate. His memberships include the International Marine Simulation Forum, International Radar and Navigation Simulator Lecturers' Conference, Maritime Training Advisory Board, the International Cargo Handling Coordination Association, and the International Maritime Lecturers Association.

J. NICHOLAS NEWMAN, NAE, is professor of naval architecture, Massachusetts Institute of Technology. He received his S.B. in naval architecture and marine engineering, S.M. in naval architecture and marine engineering, and Sc.D. in theoretical hydrodynamics from the Massachusetts Institute of Technology. Prior to joining the MIT faculty in 1967, he was a research naval architect with the David Taylor Research Center and adjunct professor, American University. His primary scientific contributions include theoretical and computational studies applicable to the interactions of ocean waves with ships and offshore platforms and wave resistance of ships. Other contributions include performance of sailing vessels, theoretical studies of maneuvering performance of ships in confined waters, and algorithms for use in navigation. He chaired the International Workshop on Water Waves and Floating Bodies in 1986 and 1988. Dr. Newman published *Marine Hydrodynamics* in 1977; he has also authored numerous other papers. He is a fellow of the Society of Naval Architects and Marine Engineers and of the Royal Institution of Naval Architects. He is a member of the American Association for the Advancement of Science and a foreign member of the Norwegian Academy of Science and Letters.

RICHARD A. SUTHERLAND, U.S. Coast Guard (retired), received his B.S. in engineering from the U.S. Coast Guard Academy. Captain Sutherland completed 30 years of Coast Guard service in 1988, as chief, Marine Safety Division in the Eighth Coast Guard District. During his Coast Guard service, he specialized in merchant marine safety and marine licensing. While serving as chief, Merchant Vessel Personnel Division, Office of Marine Safety, he managed the marine licensing program, including federal and Great Lakes pilotage, established national policies that permitted use of shiphandling simulation to gain sea-time equivalency required in marine licensing, and instituted computer-based record storage in the marine licensing program. He also headed U.S. delegations to five international meetings and served as technical advisor for seven additional meetings of the International Maritime Organization. Captain Sutherland served five years afloat in deck and engineering assignments.

APPENDIX

B

International Marine Certification Roles, Responsibilities, and Standards

National requirements for the licensing of masters and mates of seagoing ships, and professional development to meet these requirements, are based on the Standards for Training, Certification, and Watchkeeping (STCW) guidelines promulgated by the International Maritime Organization (IMO). The IMO's standards are important in evaluating ship-bridge simulation for marine training and licensing because they form the foundation for applying marine simulation to marine certification.

The IMO's recommendations for pilotage (Inter-governmental Maritime Consultative Organization Resolution A.485(XII)—Training, Qualification and Operational Procedures for Maritime Pilots Other than Deep-Sea Pilots) also serve as basic professional standards for simulation applied to certification of marine pilots. Although many of the port-level pilotage systems (administered at the state level) generally parallel the IMO's pilotage recommendations, the IMO's recommendations have not been systematically applied in the United States. The federal government has not required or encouraged state-level pilotage systems to conform to the IMO's pilotage recommendations, nor has it used these recommendations as the basis for either federal or Great Lakes pilotage systems (NRC, 1994).

This appendix describes and discusses the roles of IMO and the STCW, the IMO's recommendations for maritime pilot development, and the general adequacy of IMO standards and guidelines.

INTERNATIONAL PROFESSIONAL STANDARDS FOR MARINERS

A number of international efforts have been made to improve the safety of life and property at sea. Many of these efforts were stimulated by major marine disasters. Usually international professional standards for mariners have been preceded by national requirements. The licensing of masters and mates remains a national responsibility of maritime nations that operate a merchant marine (these nations are referred to as flag states). The licensing of marine pilots is typically administered at the port level rather than nationally (NRC, 1994).

Because mariners operate across national and international boundaries and must effectively interact with the shipping of other flag states, the qualification of mariners licensed by flag states is of interest to other flag states. Major marine accidents, especially since the *Exxon Valdez* grounding, have focused attention on improving these standards.

The qualification of marine pilots is also an important international issue, especially to the master of each vessel using pilotage services and that vessel's operating company. Most marine pilots, however, do not operate across national boundaries. Their service is highly specific to the locale in which they serve. Although pilotage has been implicated as a contributing factor in a number of marine accidents, corrective action that may be necessary has been viewed by the international maritime community as largely within the purview of the port-state pilotage authorities and has not stimulated review and improvement of international professional standards for pilots.

THE INTERNATIONAL MARITIME ORGANIZATION

Systematic international efforts to improve marine safety originated with the establishment of the Inter-governmental Maritime Consultative Organization (IMCO) as a specialized consultative organization of the United Nations. IMCO held its first meeting in 1959. The first Safety of Life at Sea (SOLAS) Convention was held in 1960. The result was the delineation of responsibilities for the safety of vessels during at-sea operations. During the 1970s, IMCO began to develop technical standards.

In 1978, IMCO promulgated the STCW guidelines. The STCW established a common standard based on marine certification practices of the traditional maritime nations, including the United States. The objective was to raise standards to a minimum level worldwide. The STCW was amended in 1991 (IMO, 1993).

In 1982, IMCO changed its name to the International Maritime Organization (IMO). The IMO is currently revising the STCW guidelines (IMO News, 1994). Much attention has been directed to strengthening STCW guidelines for the development and demonstration of competency, including the use of marine simulation. Considerable attention is being paid to the role of marine simulation in meeting STCW guidelines, and substantial guidance on using simulation is anticipated. In addition to the STCW guidelines, the IMCO also published marine pilot standards in 1981 (IMO, 1981).

IMO has relied on member countries to ratify and enforce international guidelines and standards. Generally, member countries are required to conform to the standards of the appropriate, competent international organizations. Because of accidents involving substandard ships and crews, suggesting that the actions of some flag states has been deficient, the IMO has been actively promoting port-state enforcement of IMO standards (IMO News, 1994). The IMO continues to move toward measures that would enable it to make better assessments of member states' compliance with the various international conventions.

IMO STANDARDS FOR TRAINING, CERTIFICATION, AND WATCHKEEPING

The STCW guidelines contain certification, watchkeeping, experience, and knowledge requirements for masters, chief mates, and deck officers. It also includes standards for engineering department officers and radio officers. Experience is relied on as the means by which mariners acquire practical knowledge and become capable of effectively applying theoretical and practical knowledge, skills, and abilities. A demonstration of knowledge of how to perform various functions and tasks, rather than a demonstration of ability to perform them, is the basis for basic competency determinations.

The principal exceptions with respect to determining minimum competency are contained in a resolution adopted by the 1978 International Conference on Training and Certification of Seafarers. This resolution encourages its members to require radar simulation training and training in the use of collision-avoidance aids (IMO, 1993). Experience and knowledge requirements are categorized by vessel tonnage and mariner category—master, chief mate, and officers in charge of a navigational watch. There are also mandatory minimum requirements intended to ensure the continued proficiency and updating of the knowledge of masters and all deck officers (IMO, 1993; Muirhead, 1994).

Applicability of Standards to Seagoing Ships

Tonnage categories that apply to certification and experience requirements for seagoing ships are:

- ships under 200 gross register tons,
- ships between 200 and 1,600 gross register tons, and
- ships over 1,600 gross register tons.

Marine licenses issued for operation of vessels over 1,600 gross register tons are referred to as "unlimited." For seagoing ships that are under 200 gross register tons and not engaged in near-coastal voyages, there are slightly higher requirements than for ships engaged in near-coastal voyage.

The tonnage categories that apply to knowledge requirements for seagoing ships are ships of less than 200 gross register tons and tankers and chemical ships with respect to onboard safety and pollution prevention. There are also recommendations for special training regarding vessel maneuverability by type of seagoing ship to reflect differences in handling among small ships, large ships, and vessels with unusual maneuvering characteristics (IMO, 1993).

Watchkeeping Requirements

STCW watchkeeping requirements apply to ship owners, ship operators, masters, and watchkeeping personnel—all of whom are charged to observe navigational watchkeeping procedures contained in STCW Regulation II/1. Shipmasters are specifically charged "to ensure that watchkeeping arrangements are adequate for maintaining a safe navigational watch" (IMO, 1993). Officers are responsible for the safe navigation of the vessel while on watch, under the general direction of the ship's master (IMO, 1993).

Regulation II/1 provides specific guidance on:

- watch arrangements;
- fitness for duty;
- navigation, including passage planning and frequent position fixing;
- navigational equipment;
- navigational duties and responsibilities, including watch assumption, relief, and notifications to the master;
- lookout;
- navigation with a pilot embarked; and
- protection of the marine environment.

The provisions of the regulation include areas of responsibility and overarching guidance for conducting these responsibilities. More-specific guidance on relief and conduct of the navigational watch is provided in resolutions that accompany the STCW guidelines (IMO, 1993).

Experience Requirements

There are specific service requirements for officers in charge of a navigational watch, as well as masters, and chief mates. These requirements vary by tonnage requirements and the position for which an individual is seeking certification.

Regulation II/4 specifically requires a minimum of three years of seagoing service for officers in charge of a navigational watch on ships of 200 gross register tons or more. Every candidate for such certification is required to "have seagoing service in the deck department of not less than three years which shall include at least six months of bridge watchkeeping duties under the supervision of a qualified officer" (IMO, 1993). The regulation further states that "an Administration may allow the substitution of a period of special training for not more than two years of this approved seagoing service, provided the Administration is satisfied that such training is at least equivalent in the value to the period

APPENDIX B   199

of seagoing service it replaces" (IMO, 1993). This latter provision has been interpreted by the United States and other countries as permitting the substitution of maritime education and training received by cadets at maritime academies for the required two years of seagoing service. The STCW guidelines do not specifically provide for substitution of specialized training for the required one year of seagoing service (IMO, 1993). The IMO is reviewing the adequacy of these provisions.

The STCW guidelines do not explicitly define seagoing service as service on board an actual ship. A few countries, including the United States, have interpreted seagoing service to include appropriately structured and approved marine simulator-based courses. These countries grant remission of sea service for such courses to satisfy some portions of the one-year minimum requirement for officers in charge of a navigational watch aboard ships of unlimited tonnage or seagoing service requirements for advancement. These countries generally remit a higher ratio of sea service for simulator courses than for actual service, with the ratio varying from 2 to 1 to 8 to 1 (the U.S. Coast Guard grants a 6 to 1 ratio for marine simulator-based courses that it approves for sea-time equivalency). These ratios have been arbitrarily determined based subjective determinations of value to professional development (Muirhead, 1994).

Knowledge Requirements

The STCW and its supporting regulations and resolutions contain detailed knowledge requirements. The knowledge categories for masters of unlimited tonnage vessels are outlined in Chapter 1. The knowledge categories for officers in charge of a navigation watch aboard unlimited tonnage vessels (required by an appendix to Regulation II/4) are:

- celestial navigation,
- terrestrial and coastal navigation,
- radar navigation,
- watchkeeping,
- electronic position fixing and navigation systems,
- radio direction finders and echo sounders,
- meteorology,
- magnetic and gyro compasses,
- automatic pilot,
- radio communications and visual signaling,
- fire prevention and fire-fighting appliances,
- life saving,
- emergency procedures,
- ship maneuvering and handling,
- ship stability,

- English language,
- ship construction,
- cargo handling and stowage,
- medical aid,
- search and rescue, and
- prevention of pollution of the marine environment.

There are additional, special knowledge requirements with respect to masters and officers serving aboard tankers and chemical carriers (IMO, 1993).

Recency Requirements

Regulation II/5 establishes minimum requirements to ensure the recency and updating of knowledge of masters and deck officers. At intervals not greater than five years, each ratifying members' licensing or certification authority (i.e., member administration) is required to ensure medical fitness and professional competence. The alternatives for establishing professional competence include:

- at least one year of approved seagoing service as master or deck officer within the preceding five years;
- performance of functions relevant to the duties of the certificate held, which are considered equivalent to the required seagoing service; or
- completion of one of the following: (1) passing an approved test, (2) successfully completing a prescribed course or courses, or (3) serving as a deck officer in a supernumerary capacity for a period of three months immediately prior to serving in the rank of the certificate held.

The responsible IMO administrators are also charged to:

> . . . formulate or promote the formulation of a structure of refresher and updating courses, either voluntary or mandatory, as appropriate, for masters and deck officers who are serving at sea, especially for re-entrants to seagoing service. The administration shall ensure that arrangements are made to enable all persons concerned to attend such courses as appropriate to their experience and duties. Such courses shall be approved by the administration and include changes in marine technology and relevant international regulations and recommendations concerning the safety of life at sea and the protection of the marine environment (IMO, 1993).

Simulation Training

Simulation training is strongly recommended by the STCW guidelines for all masters and deck officers, but is not required. There are no provisions to guide the specific use of marine simulation in satisfying seagoing requirements. Resolution 17 of the 1978 International Conference on Training and Certification of Seafarers, recommends "additional training for masters and chief mates of large ships and of ships with unusual maneuvering characteristics" that includes simulation training (IMO, 1993). In addition to stating that a prospective master should have suitable, general experience as master or chief mate, the resolution states that this individual "have sufficient and appropriate maneuvering experience as chief mate or supernumerary on the same ship or as master, chief mate or supernumerary on a ship having similar maneuvering characteristics . . ." or, as an alternative, "have attended an approved ship-handling simulator course on an installation capable of simulating the maneuvering characteristics of such a ship . . ." (IMO, 1993).

Updates to the STCW

The IMO is currently updating the STCW guidelines. The forthcoming revisions are expected to include a new, more definitive requirement to demonstrate a range of skills and specific performance standards. The equivalency articles with respect to seagoing services are expected to remain in place. The one-year minimum seagoing service requirement may be redefined as one year of actual onboard experience, although there may not be an actual prohibition against using simulation to satisfy a portion of the one-year requirement. Extensive policy and technical guidance is being developed for the application of marine simulation.

IMO RECOMMENDATIONS FOR MARITIME PILOTS

IMCO Resolution A.485 (XII)—Training, Qualification and Operational Procedures for Maritime Pilots Other than Deep-Sea Pilots—provides recommended standards for the certification and qualification of maritime pilots who are:

- not deep-sea pilots (pilots providing pilotage services outside of a local pilotage area);
- ship's masters; or
- members of a ship's crew who are authorized by license or certificate to perform pilotage duties in specific areas by a competent authority (that is, the responsible administration within an IMO member country or another entity that by law or tradition provides a pilotage service).

The competent authority may exempt from the provisions of the resolution any individual who carries out only berthing duties.

The resolution calls for each maritime pilot to:

- be medically fit;
- meet STCW standards for certification of masters, chief mates, and officers in charge of a navigational watch;
- be appropriately certificated or licensed; and
- meet the additional standards of the resolution.

It also encourages the certificate or license to indicate the extent of a pilot's authority to provide services, such as maximum size, draft, or tonnage. With respect to continued proficiency and the updating of knowledge, the resolution states that the competent authority must satisfy itself that, at not greater than five-year intervals and after extended absences from service, all pilots for which it exercises jurisdiction:

- continue to possess recent navigational knowledge of the local area to which the certificate or license applies;
- continue to meet the medical fitness standards [prescribed in the resolution];
- possess knowledge of the current international, national, and local laws, regulations, and other requirements and provisions relevant to the pilotage area or duties (IMO, 1981).

The resolution also recommends that each competent authority have control over the training and certification of pilots, including:

- the development of standards;
- administration of prerequisites for training, examination, and the issuance of certificates or licenses; and
- the investigation of incidents involving the service of pilots (IMO, 1981).

Recommended knowledge requirements for maritime pilots include:

- boundaries of local pilotage service areas;
- the International Rules for the Preventing of Collisions at Sea and applicable national and local safety and pollution prevention rules;
- systems of buoyage in use in the pilotage service area;
- characteristics of local lighted aids to navigation, fog signals, and beacons;
- all relevant information about other aids to navigation in the pilotage service area;
- pertinent channel, geographic, and topographic data;
- proper courses and distances;
- traffic separation and routing schemes, ships' services, and traffic management systems (i.e., vessel traffic services);
- hydrographic data, including tidal and current effects;
- anchorages;
- ship-bridge equipment and other navigational aids;
- radar and automatic radar plotting aids;
- communications and available navigational information;
- radio navigational warning broadcast systems;
- maneuvering characteristics for all vessels piloted, including any limitations associated with various propulsion and steering systems;
- hydrodynamic and other physical factors affecting ship performance;
- assist tugs;

- English language skills, with sufficient fluency to express communications clearly; and
- other knowledge considered necessary by the competent authority (IMO, 1981).

ADEQUACY OF INTERNATIONAL STANDARDS

To obtain the support of IMO member states for their acceptance and implementation, the STCW standards reflect a middle ground among all the existing national requirements. The STCW forms modest baseline standards. Its requirements with respect to knowledge, skills, and abilities are based on longstanding traditional practices known to work if effectively applied. The adequacy of these standards with respect to modern navigation technology has not been assessed. Furthermore, there are no content or quality control requirements to guide the development of knowledge, skills, and abilities through sea service, nor have sea-time requirements been validated through a program of research to determine their relevancy and adequacy.

Flag-state licensing authorities had few motivations to exceed STCW requirements, although forward-looking shipping companies and unions often have programs that exceed both applicable license criteria and STCW provisions. That the STCW is undergoing considerable updating reflects increased interest in improving marine safety as a result of major marine accidents with extensive pollution, beginning with the grounding of the *Exxon Valdez* in 1989. Action to update the STCW guidelines also reflects flag-state and marine industry concerns about industry trends to reduce crew sizes and operating costs, and the safety implications of resulting practices, such as one-man watchkeeping at night.

REFERENCES

IMO (International Maritime Organization). 1981. Training, Qualification and Operational Procedures for Maritime Pilots Other than Deep-Sea Pilots. IMCO Resolution A.485(XII) adopted on 19 November 1981. London, England: IMO.

IMO (International Maritime Organization). 1993. STCW 1978: International Convention on Standards of Training, Certification, and Watchkeeping, 1978. London, England: IMO.

IMO News. 1994. World maritime day 1994: better standards, training and certification—IMO's response to human error. IMO News (3):i-xii.

Muirhead, P. 1994. World Maritime University, personal communication to Wayne Young, Marine Board, National Research Council, September 20.

NRC (National Research Council). 1994. Minding the Helm: Marine Navigation and Piloting. Committee on Advances in Navigation and Piloting, Marine Board. Washington, D.C.: National Academy Press.

APPENDIX

C

Professional Licensing Infrastructure for U.S. Merchant Mariners

U.S. COAST GUARD MARINE LICENSING PROGRAM

Competency Determinations Through Essay and Multiple-Choice Examinations

From 1942 until the mid-1970s, the U.S. Coast Guard's (USCG) license test consisted of essay examinations. The licensing administration system consisted of hundreds of examination cards, with different questions for each major topic. Each marine inspection officer was provided the same set of cards, containing a separate question. Some cards were designed for individual license levels, others applied to all license levels. The licensing examiner would prepare an individualized test for each person by pulling cards from the file. Thus, each test was different. The consistency of the examinations varied with the licensing examiner's ability to organize and correlate question content, level, and difficulty and attitude toward the individual being tested Examinations also varied among licensing examiners. The number of questions given varied from 5 to about 20, depending on the topic.

The licensing examiner was empowered to use his or her professional expertise to evaluate the license candidate's response. Options available to the examiners were to:

- fail the individual,
- give the candidate additional questions to provide a broader base for evaluation, or
- prepare an alternate response if it was clear that the candidate did not understand the question.

This approach was necessary, in part, because many questions retained in the updated system were prepared before World War II. Licensing examiners were often individuals who had sailed in the merchant marine and who held deck or engineer merchant marine licenses and had the professional expertise needed to evaluate subjective responses to licensing exam questions.

In 1953, the USCG conducted a major study of the examination process. In addition to the unevenness and obsolescence previously noted, the agency found that the essay system had been compromised. Wohlfarth (1978) found that:

> This compromise occurred with the compilation of "ponies" which accurately reproduced license examination questions. Allegedly some were exact duplicates of the card files maintained by the examiners.
>
> A part of the compromise problem was the tendency of some schools to teach from such ponies, and the inclination of some license applicants to learn solely from ponies by rote memorization.
>
> The U.S. Coast Guard was partially responsible for the existence of such ponies through its failure to conduct timely revisions to eliminate obsolete, ambiguous, and incorrect questions. Some license applicants felt compelled to study such ponies in order to learn "the answers required by the U.S. Coast Guard," thus enhancing their chances of passing the examination.

Although the 1953 study did not lead to direct changes, its findings indirectly resulted in the inclusion of multiple-choice questions for the first time in license examinations. Although there are several exceptions, the USCG ultimately changed its license program to rely almost exclusively on multiple-choice examinations to ascertain an individual's competence. One exception is a requirement that candidates for pilot licenses or endorsements draw a chart from memory of the route for which pilotage is sought. Candidates for master, mate, and pilot licenses must also complete USCG-approved training leading to issuance of a radar observer certificate. With these exceptions, there are no federal marine license requirements that incorporate a demonstration of competency.

The shift to the multiple-choice format began in 1969. Education Testing Services (ETS) in Princeton, New Jersey, was contracted to study all aspects of the marine licensing program. The study resulted in recommendations for interim and long-term improvements. In response to the study, the USCG:

- developed job descriptions for masters and mates, and
- eliminated essay questions from the license examinations, substituting multiple-choice questions.

The ETS was contracted to prepare the job descriptions and to develop multiple-choice examinations.

The job descriptions were developed in broad consultation with the marine community. Consultants were subcontracted to write the multiple-choice questions, which were pretested at some of the maritime academies, union schools, and USCG marine inspection offices. Acceptable questions became part of printed examinations and were provided in early 1973 to a newly established division at the USCG Institute (Wohlfarth, 1978). The Institute itself had been relocated to Oklahoma City, partly to take advantage of mainframe computer resources that could be used for scoring the license examinations.

There were significant problems in the revised licensing program during implementation, which were corrected as experience was gained with the new format. The multiple-choice format was generally accepted within the marine community. This initial success was attributed to the extensive coordination and consultations that were conducted (Wohlfarth, 1978). Dissatisfaction was limited principally to specific questions rather than to the format. Mariners and operating companies also benefited from the substantially reduced time needed to complete the examination. It is unclear how much this contributed to acceptance of the new format, although it was probably a significant factor.

In recent years, the USCG's multiple-choice examination approach has been criticized widely within the marine and towing industries and by marine pilots. One concern is that the examination does not ensure that an individual is professionally competent. Because all the examination questions were released by the USCG to the public in response to Freedom of Information Act requests, the integrity of the examination process has been compromised. This public release and the advent of computer-aided teaming techniques has enabled the creation of "electronic ponies," which are in wide use at many license-preparation schools. The rote memorization of the essay examination of the past has been replaced by the "programming" of license candidates to successfully pass license examinations.

Dissatisfaction with the multiple-choice examination format has caused some marine groups to call for a return to the essay format. Such an approach would be confronted by the same types of problems that the earlier essay system experienced. One alternative is to involve marine simulation in the process, either as required training, as an evaluation platform, or as a combination of these two options.

Nautical Credentials of USCG Marine Licensing Officials

The nautical expertise of marine licensing officials is an important consideration in applying marine simulation to professional regulation. The marine licensing authority must either maintain adequate resident expertise to fully administer the program or find alternative sources of expertise for the same purpose. It would be difficult for the USCG to revert to the earlier system of licensing, because the agency has not maintained a cadre of licensing examiners with the requisite qualifications. As discussed in the main body of this report, a move by the USCG to use simulators could also necessitate the maintenance of a staff with extensive marine qualifications to effectively oversee the validation and use of marine simulation.

The merchant marine and towing industry operational experience and expertise possessed by USCG personnel involved in marine licensing have been criticized by the marine industry and the piloting profession as inadequate. Members of the industry have also criticized the agency's capability to establish professional standards for commercial mariners and to credibly evaluate mariner knowledge, skills, and abilities. This latter complaint was not a problem in 1972, when the multiple-choice examination format was established. Indeed, the individuals assigned to the USCG Institute to administer and update the examination questions all had considerable prior seagoing service, and all but one held merchant marine licenses (Wohlfarth, 1978).

Over the past two decades, the USCG's basis for establishing seagoing professional expertise among its officer and enlisted corps has progressively deteriorated (NRC, 1994). Generally, the seagoing expertise of most USCG members is derived from USCG cutters, most of which are designed for naval service. The operations and shipboard organization aboard cutters are substantially different from those aboard commercial ships and towing vessels. Thus, a substantial portion of nautical experience of some Coastguardsmen is of limited relevance to commercial operations.

Determining the actual status of USCG nautical expertise is not easily accomplished because the agency does not maintain automated databases that would permit the agency to determine the nautical qualifications of the merchant marine and sea service of individuals who conduct marine licensing at regional examination centers. An informal survey of 36 military personnel and civilian employees that administer the marine licensing program at USCG Headquarters in Washington, D.C., determined that 23 percent had prior sea-service experience in either the USCG or the merchant marine and 17 held merchant marine licenses. The number of individuals holding marine licenses at the agency's regional licensing examination centers is substantially less (NRC, 1994).

USCG Assessment of Licensing Alternatives

The USCG is acutely aware of the criticism of its marine licensing program and the nautical credentials of marine licensing personnel. The agency is interested in ways to incorporate competency determinations in the professional regulation of mariners and has been active in advancing the concept of using marine simulation for this purpose nationally and through its international representation of U.S. interests before the International Maritime Organization and its committees and working groups.

The USCG conducted an internal assessment of its marine licensing program as a partial response to public and congressional criticism of the professional qualification of mariners. The resulting report, *Licensing 2000 and Beyond* (Anderson et al., 1993), covers a broad array of licensing issues.

The report was prepared by a focus group consisting entirely of USCG uniformed armed service and civilian personnel and reflects their perspective. Representatives of user groups were not included in the focus group membership.

The focus group recognized its composition and urged that its findings and recommendations be referred to and validated by the USCG's new Merchant Personnel Advisory Committee (MERPAC) and USCG's Merchant Vessel Personnel Division at USCG Headquarters in Washington, D.C. The report also contains recommendations within the areas of interest of the USCG's Navigation Safety Advisory Committee and Towing Industry Advisory Committee.

The fundamental nature of changes recommended by the report are far reaching with respect to their potential effects on the merchant marine. The report provides a conceptual basis for guiding improvements in marine licensing, but its supporting facts and analyses are limited. The focus group acknowledged that its resources were insufficient to adequately develop some important inquiries. The resulting facts and analyses provide only a partial basis for identifying and assessing the potential effects.

With respect to USCG-approved courses, the report states that "proper utilization of the approved course concept would allow reductions in actual sea service experience, enhancement of professional proficiency and in many instances elimination of the U.S. Coast Guard Examination Process." The assertion that sea service can be reduced and some USCG examination requirements eliminated are not substantiated by facts and analyses. In particular, no substantive basis is provided to justify the levels of substitution of sea time.

The requirement for radar observer certification (which uses radar simulation) is cited as a working example of the use of simulation, but no evidence is presented, nor does there appear to be any evidence that the training is actually improving the licensing process, or—more important—improving the quality of professional development for effective use of radar. Radar observer training, as it is presently used, has been criticized by many within the marine community and piloting profession because the requirement has not kept pace with the change in radar technology, especially with respect to automated radar plotting aids. Furthermore, the effectiveness of radar observer certification has not been evaluated empirically with respect to knowledge and skill requirements, the degree to which these skills are reinforced or refreshed during actual operations, and the frequency and nature of part-task training that is necessary to restore degraded knowledge and skills.

The report states that improvements are needed in the qualification of mariners and instructors and in the preparation of USCG personnel for administrative oversight of agency-approved courses. The need for instructor qualifications is a view that is widely shared within the marine community and the piloting profession. The report also strongly encourages that USCG personnel involved in marine licensing receive more complete and rigorous training in the licensing process. The report is silent, however, with regard to the nautical expertise and recency that would be needed for effective performance by USCG personnel who serve as licensing examiners, who set marine licensing standards, or who

APPENDIX C 209

validate USCG-approved courses. It implies that expertise can be developed through assignments of personnel in the vessel inspection program.

The report states that use of marine simulation for testing purposes is controversial and that "the wide-spread use of simulation as a test of more definitive subjective knowledge has yet to be fully demonstrated" (Anderson et al., 1993). It characterizes the use of marine simulation by operating companies as successful, based on use of the technology rather than on the practical results of the training in actual operations. Although the companies and marine pilot associations that use simulation believe that there is value to their operations, and thus continue to use simulation, the value added has not been determined empirically.

The report recommends that the potential for applying simulation in training and licensing be assessed and recommends the adoption of simulation for demonstrations of competency. In particular, the report recommends development of "performance standards for a high current/tight quarters maneuvering simulator training program" and a requirement to complete such a "simulator training and testing program as a prerequisite for issuance of a Western Rivers OUTV." The state of practice in computer-based simulation of inland towboat operations, especially for large barge flotillas, is not identified. The training resources that are currently available to support the training of inland towing vessel operators is the same infrastructure that is used for masters, mates, and pilots. For these reasons, it is not clear that the focus group's recommendations could be implemented in the near term or mid-term.

The report defined competency as:

> ... the total set of skills, knowledge and judgments necessary for the proper performance of one's duties in a specific position on a specific vessel. Competency, thus defined, can be broken down into two subsets. The first is the base level of skills, knowledge and judgments necessary to perform the duties of generic positions, e.g., Chief Mate, on a wide range of vessels of similar size and type. ... The second competency subset consists of the skills, knowledge and judgments peculiar to a specific vessel or trade (Anderson et al., 1993).

The focus group in its report further stated that:

> Ensuring that mariners possess the base level, also known as minimum competence, is a proper role of government. The consensus of the Focus Group is that public safety and the environment will be better protected by improving the methods by which mariners obtain this minimum competence. The second competency subset consists of the skills, knowledge and judgments peculiar to a specific vessel or trade. It is the responsibility of the owners, operators and individual mariners to ensure that these are obtained in a timely manner (Anderson et al., 1993).

The report found that the level of improvement varied among the elements of the marine licensing program. The focus group recommended that the USCG move forward in assessing computer-based training and testing systems and in using these methods to verify professional competency.

The USCG has not publicly stated its official position on the report's findings, having recently referred it to MERPAC for validation. Nevertheless, the USCG has proceeded on its own initiative to advance the application of marine simulation in professional development and marine licensing. Many courses had previously been approved for sea-time credit (USCG, 1994). In November 1994, the agency published its course-approval policy guidance, which strongly encourages simulation training and provides general criteria for the granting of sea-time equivalency. (Subpart C of 46 CFR 10 contains the general criteria used by the USCG to approve courses of instruction.) Also in November 1994, the agency established a precedent by accepting a combined simulation-based training and evaluation course as a nonmandatory substitute for the entire master (unlimited oceans) license examination (see Appendix F).

REFERENCES

Anderson, D.B., T.L. Rice, R.G. Ross, J.D. Pendergraft, C.D. Kakuska, D.F. Meyers, S.J. Szczepaniak, and P.A. Stutman. 1993. Licensing 2000 and Beyond. Washington, D.C.: Office of Marine Safety, Security, and Environmental Protection, U.S. Coast Guard.

NRC (National Research Council). 1994. Minding the Helm: Marine Navigation and Piloting. Committee on Advances in Navigation and Piloting, Marine Board. Washington, D.C.: National Academy Press.

USCG (U.S. Coast Guard). 1994. U.S. Coast Guard approved course listing as of October 1, 1994. Unpublished. Merchant Personnel Division, Office of Marine Safety, Security, and Environmental Protection, U.S. Coast Guard, Washington, D.C.

Wohlfarth, W.G. 1978. Licensing examination modernization. Pp. 4–9 in Proceedings of the Marine Safety Council, January. Washington, D.C.: U.S. Coast Guard.

APPENDIX
D

Hydrodynamics, Physical Models, and Mathematical Modeling[1]

Marine simulation for channel design and mariner training developed along two parallel and complementary lines—physical scale models and computer-based simulations. The first computer-based simulators were based on mathematical models for a ship's dynamics in deep, unrestricted waters. These early models were coupled with rudimentary bridge mockups controlled by computers. Simulation technology has evolved with improvements in computer hardware, computer-generated imagery (for ship-bridge simulations), and increasing knowledge from naval architects of the appropriate models for ship dynamics.

Beyond ship dynamics, modeling of other operating conditions—including shallow-water effects; restricted-water effects, including bank-suction; slow-speed maneuvering, including anchoring and normal backing evolutions; turning with tugs and thrusters; and ship-to-ship interactions—is more complex than mathematical modeling for ships operating in deep and unrestricted shallow water. The state of knowledge in these areas is technically less advanced. In addition, the modeling of vessels towing barges alongside on a line, or pushed ahead, and the modeling of integrated tug and barge units have only recently received the level of attention formerly given to ship dynamics modeling. Modeling of towing vessels can be more technically complicated than for ships in general.

Complementary developments to ship-bridge simulation capabilities have occurred with the use of physical scale models of ships. The principal use has

---

[1] As input for the work of the Committee on Ship-Bridge Simulation Training, the Society of Naval Architects and Marine Engineers, Panel H-10 (Ship Controllability), held a workshop on the role of hydrodynamics in ship-bridge simulator training (SNAME, 1993).

been as manned scale models of ships, referred to as manned models, used primarily for shiphandling training. Radio-controlled scale models have also been used for shiphandling training, but only to a very limited extent. Although scale models have not been developed for training in either the coastwise or inland towing industries, where on-the-job training for shiphandling is common practice, they have been used extensively in channel design and to develop maneuvering strategies in new and unusual situations.

The level of trajectory and rates of motion (i.e., vessel response) accuracy and fidelity that are needed and delivered in the replication of ship maneuvering behavior for simulation training in both computer-based and manned-model simulation are debated within the hydrodynamic modeling, marine simulation, and marine education and training communities.

This appendix describes the modeling of vessel maneuvering used in marine simulation, the levels of accuracy present in the various modeling approaches, and where the different modeling approaches and levels of sophistication and accuracy may be most appropriate for specific training purposes. Accuracy and fidelity are also discussed from a modeling viewpoint. What constitutes a vessel's inherent maneuvering capability are described, including how these capabilities affect operations. A brief description is given of the development of modeling and simulation for marine training, with some comments on utility. Future developments are discussed, along with their potential for practical implementation. The final section is a general summary of modeling approaches and levels of sophistication available to support training objectives.

GENERAL DEVELOPMENT OF MARINE SIMULATION

Computer-Based Simulation

Ship maneuvering, a branch of naval architecture, originated from the need to design ships with maneuverability characteristics that either meet specific requirements (turning circle diameter, or tactical diameter, was an early specified requirement) or are reasonable for the mission of the ship. As the mathematical theory and the hydrodynamics of ship movement advanced, more accurate, computer-driven mathematical models were developed to represent and predict ships' trajectories.

Although analog computers were used for early models, digital computer modeling replaced analog because of the complexity of the models used for ship design. Shiphandling simulators capable of involving the human in a real-time experience were developed by combining digital computer-based models with bridge equipment, bridge mockups, and visual projection systems. As computer technology and computer-generated images advanced, so to did the shiphandling simulator (Puglisi, 1987). Modern computers made it practical to create ship-bridge simulators for full-mission and multi-mission training. Computers also

APPENDIX D 213

made it practical to combine actual radar equipment with mathematical models of vessel behavior to create radar simulators for use as an element of full-mission or multi-task training or as a limited-task or special-task stand-alone training device.

Full-mission, multi-task, and limited-task simulators are, as a rule, operated in real time[2] and can appear to be highly realistic. The amount of realism is referred to as "face" or "apparent" validity (NRC, 1992).[3] Ship-bridge simulators are used for all types of operational scenarios. Important issues in the use of computer-based marine simulation include:

- whether all of the appropriate vessel maneuverability cues are present in the simulation or correctly portrayed,
- whether the maneuvering response of the ship is actually correct, and
- the relative importance of accuracy in these areas (NRC, 1992).

Development of ship-bridge simulators and simulations is more complex than development of commercial air carrier simulators. Development of visual flight simulators for the commercial air carrier industry is linked directly to the development of specific airframes. The simulators are not modified to permit training on multiple airframes (NRC, 1992). This practice is possible because of the large numbers of similar airframes owned and operated by commercial airlines.

Ship-bridge simulators are developed independent of the vessels they simulate and are routinely adjusted to permit training in other hull forms and sizes. As a result, some simulator facilities use either a number of models to meet the specific application needs of training sponsors or adjust their model to simulate a different type or size vessel. If these adjustments are not correct, the resulting trajectory predictions are inaccurate, regardless of the quality of the algorithms used or the apparent validity of the simulation. For these reasons, it is appropriate to validate each trajectory prediction model or perturbation in a model to determine the capabilities and limitations of the product being delivered to the trainer, the marine licensing authority, and licensing examiners and assessors.

Radar Simulators

Radar simulators are an example of effective use of limited-task simulation for mariner training. Radar simulators, first used for mariner training in the 1960s, were developed separately from ship-bridge simulators. They used

---

[2]Combined real-time and fast-time computer-based simulations have been used outside the commercial maritime sector, for example, to minimize the delay between planned learning scenarios.

[3]Sometimes training simulations are run in fast time to bring trainees quickly to learning situations, which are then run in real time, such as surface warfare simulations. This approach has not been common with the use of ship-bridge simulators for training merchant mariners.

analog computers and coastline generators to generate visual presentations on actual radar equipment. Simple linear equations of motion were used for ship trajectory predictions. Although digital coastline generators were available as early as 1973 for military applications, transition to digital radar simulators in commercial marine applications followed the introduction of digital radars into commercial maritime operations in the early 1980s. Today, virtually all radar simulators use digital data.

The development of digital computers allowed the use of sophisticated, mathematical trajectory and prediction models to drive radar simulations, either independently or as an element of a ship-bridge simulation. The equations of motion used in stand-alone radar simulators are generally less sophisticated than those used in ship-bridge simulators, because of cost and because a high level of trajectory accuracy is not considered necessary for basic radar, limited-task training exercises.

Radar simulators are operated in real time. When operated as stand-alone simulators, the primary training objective is generally radar plotting rather than watchkeeping or piloting. Understanding the capabilities of the equation of motions used to drive the radar simulation is essential for determining whether a particular radar simulator is suitable for navigation and piloting training. This understanding is especially important where shallow- or restricted-water effects are necessary to achieving training objectives and must be simulated with greater trajectory accuracy than might otherwise be possible with some radar simulators.

Manned Ship Models

Physical modeling of ships was first applied to training in 1966, with the building of the world's first manned-model training facility in France. This facility was initially developed for training ESSO Marine's masters in the maneuvering capabilities and shiphandling procedures for very large crude carriers. At that time, very large crude carriers represented a quantum jump in vessel size, with attendant changes in maneuverability from conventional ships. An engineering organization with extensive experience in port and harbor development was selected as the contractor and has operated the facility commercially since its opening (Graff, 1988). Three additional manned-model training facilities have been developed, one in England, in the United States (at the U.S. Navy Amphibious Base in Little Creek, Virginia), and most recently in Poland. The Little Creek facility, where a number of merchant mariners received shiphandling training in conjunction with Naval Reserve training, was closed in 1993 by the U.S. Navy.

Physical models, in contrast to ship-bridge and radar simulators, always simulate ship motions and shiphandling in fast time because of scaling factors. Manned models are believed by many to provide realistic representation of bank effects, shallow water, and ship-to-ship interactions. The manned-model hull forms and

APPENDIX D

water medium result in "automatic" representation of the hydrodynamic forces acting on ships during typical maneuvers for the particular ships modeled. Limitations of scale models include the exaggerated effect of wind, restrictions on the number of different ship types and channel configurations, stereoscopic effects of human perception at a reduced scale, and the significantly compressed time scale of operations. In particular, the individual must adjust to vessel size and time scaling during training and interpolate and assimilate the effects to subsequently apply the knowledge, skills, and abilities in actual operations.

Modeling of Towing Vessels

The modeling of vessels that tow barges—either by pushing a flotilla, pulling a barge attached alongside, or pulling a barge on a line—is different from modeling a more typical ship design. The focus of this discussion is on pushtows, which are the dominant type operating on U.S. inland waterways.

For pushtows, the basic hull form is dramatically different than a ship and complicates the flow patterns and pressure fields around the hull. The shape is very square and flat, with a relatively blunt bow and stern shape causing severe flow separation. The power unit, with the propellers and rudders, is often relatively small in beam, and there are usually two or more propellers for thrust. Because river operations require significant maneuvering and backing capabilities, towboats are equipped with multiple rudder sets (one in front of the propellers and one behind the propellers) for flanking and steering operations, respectively. Each propeller and rudder set is independently controlled and allows the tow to be operated in modes that are very complex to model (e.g., one propeller operating ahead and one in reverse, both flanking and steering rudders placed to turn the vessel in opposite directions, and the tow either moving ahead or astern). Towing-tank tests show that behavior in shallow water changes from that of finer-formed ships. The flotilla of tows is constantly changing, with barges being added and dropped off during transits.

These flotillas are lashed together with sets of ropes, "wires," or cables, creating a semirigid unit. Tug/barge units with the tow lashed alongside have a very asymmetric shape, with the thrust being applied so that there are strong moments about the unit's center of gravity. Tugs pulling a barge on a line are essentially a multibody modeling problem, with each unit having its own hydrodynamic model, and forces and moments being applied via the line connecting them. Except for the latter case, tug/barge units are typically modeled as rigid units, assuming that they are tightly lashed.

FIDELITY AND ACCURACY

For training purposes, it is important to provide the necessary level of realism and accuracy to support training objectives. Defining the necessary level of fidelity is important because cost generally increases with the level of realism and accuracy provided.

The Fidelity-Accuracy Relationship

Fidelity refers to the realism, or degree of similarity, between the training situation and the operational situation being simulated. The two basic measures of fidelity are physical and functional characteristics of the training situation (Hays and Singer, 1989). In the case of a manned model, the model contributes to both the physical and functional characteristics of the simulation. In the case of computer-based simulation, the mathematical model contributes to the functional characteristics of the simulation. Fidelity is determined subjectively. The level of fidelity required to achieve training objectives should be based on task needs and training analysis (Hays and Singer, 1989).

Determinant measures may be used to aid in assessing the level of fidelity present in a given simulation. Accuracy is inherently a determinant measure of how close something is to being exact. The accuracy of a maneuvering model is typically determined by measuring variations of the predicted trajectory with the actual trajectory for a given set of controls and environmental operating conditions.

Correlating Realism and Accuracy

In many respects, fidelity is more difficult to address than accuracy, because fidelity involves a subjective assessment of how real the simulation is. Balancing accuracy of trajectory modeling with fidelity of motion in visual scenes, for example, is very challenging. It is possible to provide a believable simulation using a simple trajectory model that, with a few minor validating adjustments, can appear to be realistic to pilots and mariners in a specific harbor and ship situation. Performing different maneuvers than those used for validation, however, can result in quite inaccurate trajectories. All models have limitations in accuracy in various regimes that may not always be avoided by the trainer who may be unaware of these limitations. In general, this issue has not been addressed by simulation providers except to try to use the most accurate modeling approach economically available to them. Given the state of modeling practice, this approach usually results in acceptable modeling.

The accuracy of the trajectory prediction models available to drive a simulation can be compared with the level of fidelity specified by the training analysis as necessary to achieve training objectives. The accuracy of trajectory prediction, for instance, would have lesser importance in courses in which vessel maneuvering behavior was not an instructional objective than in courses in which maneuvering was required to achieve the goal of certain learning situations or was the primary instructional objective.

Deliberate Departure from Realism

It is sometimes possible to enhance training effectiveness by departing from realism. As a general rule, however, in marine simulation, departures from realism are driven by limitations in training resources rather than a conscious attempt to optimize training effectiveness. The most notable exception is the initial development of manned models, a development borne out of practical necessity to safely train the prospective masters of very large crude carriers in shiphandling, which predates the research-psychology literature on training fidelity (Hays and Singer, 1989). As discussed in this appendix, the scaling inherent in manned models is believed by many to enhance training effectiveness, although there are concerns about the effect of scaling factors on individuals who do not have a well-established frame of reference in the operation of ships of the categories being simulated.

Because computer-based simulations rely primarily on software-based mathematical algorithms, there is considerable flexibility that could be used to deliberately depart from realism. In marine simulation, however, the opposite approach has been the rule. Most simulator facilities have sought high degrees of realism to build and improve confidence in simulation capabilities among mariners, training sponsors, and marine licensing authorities.

It is possible to alter the mathematical trajectory prediction models to accentuate certain vessel maneuvering behavior, for example, as an instruction technique to assist a trainee in becoming aware of a particular behavior. As a rule, such an approach is problematic, because it appears that only a few ship-bridge simulation staffs have reached the level of sophistication in instructional design and hydrodynamic modeling to effectively stage and control deliberate departures from realism. As discussed in a later section, there are strong reasons, from a hydrodynamic perspective, to avoid field adjustments to trajectory prediction models.

ELEMENTS OF SHIP MANEUVERABILITY

To understand how simulation modeling techniques differ and to better judge the level of trajectory modeling accuracy necessary for a specific training exercise, it is necessary to understand the phenomena that are being modeled. There are several primary areas into which a ship's maneuvering capabilities can be categorized:

- turning ability,
- checking ability (recovery from a turn),
- course-keeping ability,
- stopping and backing ability, and
- operability at slow speeds.

A vessel's capabilities at different trims and loading conditions, as well as in shallow and restricted waters (channels, banks, and other constrictions), change significantly from those in deep water and in a complex fashion. This section begins by describing the operational results of maneuvering capabilities in deep water. In general, much greater sophistication is required to accurately model trajectories than at other conditions.

Turning, Checking, and Course-Keeping Abilities

Turning, turn recovery, and course-keeping abilities are closely related to the level of dynamic course stability, a characteristic of hull form and rudder. In general terms, dynamic stability is the ability of a vessel to return to a steady heading (or initial turning condition) after a disturbance. Figure D-1, for example, shows the response of stable and unstable ships with the rudder fixed, after a

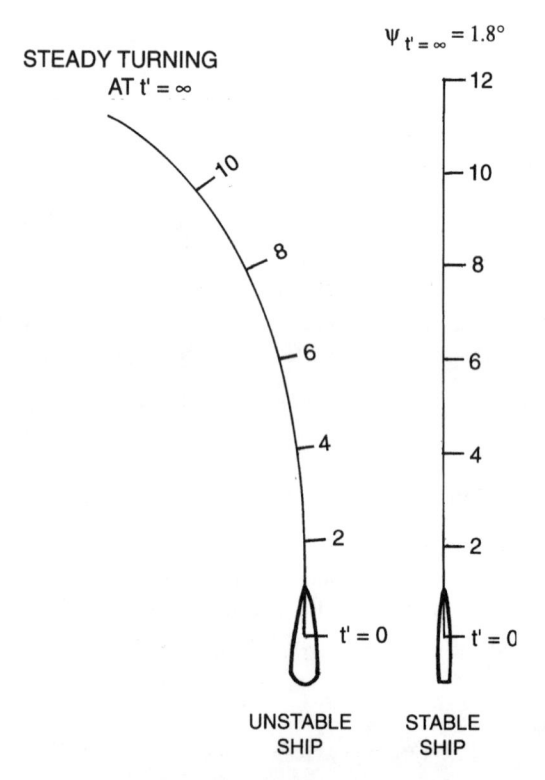

FIGURE D-1 Paths of stable and unstable ships after a yaw disturbance of 1 degree (t' = ship length of travel). Source: Eda and Landsburg (1983).

TABLE D-1 Principal Particulars

	Hull Form		
	A Slender Fine	B Wide Beam Full	C Very Wide Full
Stability	Stable	Unstable	Very Unstable
Length/Beam	6.95	5.0	4.0
Length/Draft	19.56	16.2	16.2
Beam/Draft	2.81	3.24	4.03
Block Coefficient	0.613	0.820	0.810
Prismatic Coefficient	0.625	0.823	0.813
LCG^a fwd ⊗ /L	−0.015	0.026	0.019
Rudder Area/Length x Draft	1/45	1/53	1/48

[a]LCG = longitudinal center of gravity. This is the location of LCG forward of midship/ship length.
Source: Eda and Landsburg (1983).

yaw disturbance of, say, 1 degree is given from an initial straight course (Eda and Landsburg, 1983). The stable ship eventually comes to a new straight course, which is near the original heading. The heading of an unstable vessel, on the other hand, continues to change with time, until nonlinear hydrodynamic forces override the inherent instability of the hull form. Such a ship requires constant helm corrections to maintain a desired heading.

Many ships have been built with some degree of inherent instability. Typical examples are large tankers, which generally have inherent instability of course at loaded conditions because of relatively large values of the block coefficient and beam-to-length ratios. The degree of instability is substantially increased with an increase in beam-to-draft ratio, which is the case for shallow-draft, wide-beam ships.

The practical effects of dynamic instability can be understood by reviewing trajectory results from three ships with different levels of dynamic instability (Eda and Landsburg, 1983). The type and characteristics of the hulls are given in Table D-1. The solid-line curves in Figure D-2 show steady turning rates (predicted results from Figure D-3 spiral tests) for ships A, B, and C. Arrows along the curves show the sequence of results predicted for the spiral tests. Dotted lines indicate the jump in steady turning rates during spiral tests of dynamically unstable ships B and C. Predicted zig-zag maneuver (Z maneuvers) trajectories were computed for these ships at an approach speed of 14.5 knots.

Results of the simulations for ships A, B, and C (Figure D-4) indicate dynamic behavior during the Z maneuvers. The dynamically stable ship, A, has a small overshoot angle, and can quickly finish a Z maneuver. The unstable ship, B, has a larger overshoot angle, and it takes more time to complete the test than

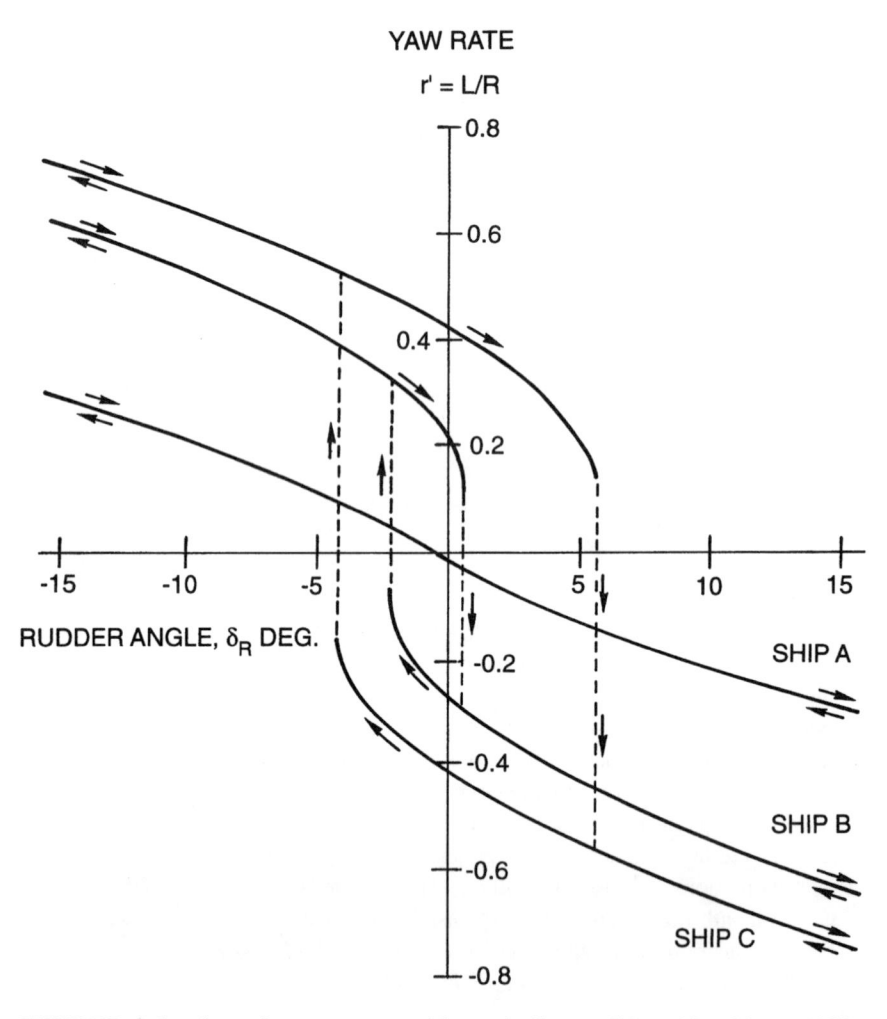

FIGURE D-2 Steady turning rate versus rudder angle. Source: Eda and Landsburg (1983)

ship A which is stable. While ship B finished the 15–15 degree Z test in a stable fashion, it could not finish Z maneuver tests of 7.5–7.5 and 5–5 degrees in a stable manner (i.e., heading angle is oscillatory divergent in unstable patterns).

In the case of the very unstable ship, C, where the heading angle is divergent after the first execution of the rudder angle, recovery cannot be achieved by the use of the opposing rudder angle at the second execution. The ship did finish the 15–15 degree Z maneuver test in stable fashion because of the significant contribution of the nonlinear terms.

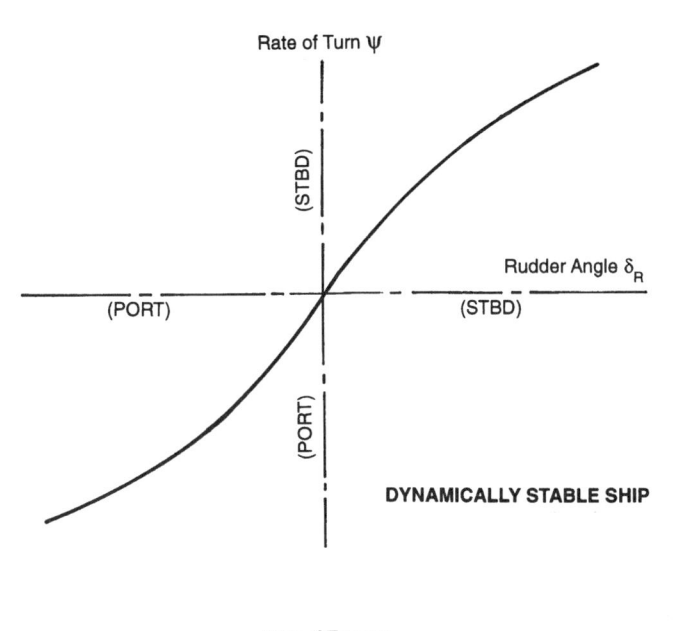

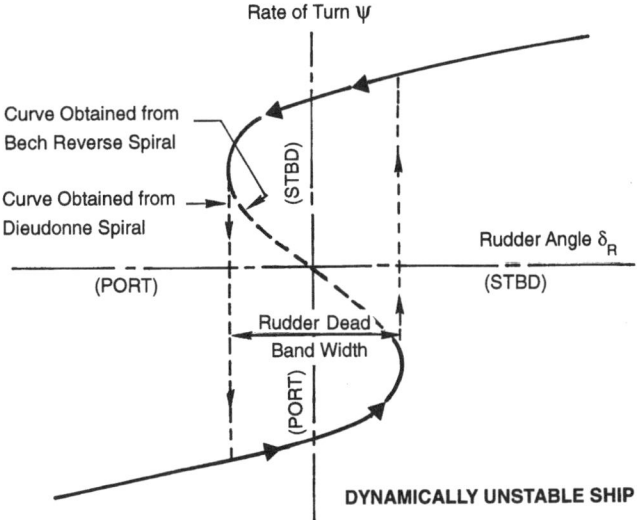

FIGURE D-3 Spiral test. Source: Adapted from Crane et al. (1989).

Operations in shallow water, in restricted water (including banks), and interactions with other ships also produce significant changes in the dynamic course stability of the hull, which in turn results in different maneuvering behaviors. Vessel draft, and especially trim, are also critical parameters affecting dynamic course stability. Physical models automatically account for these behaviors,

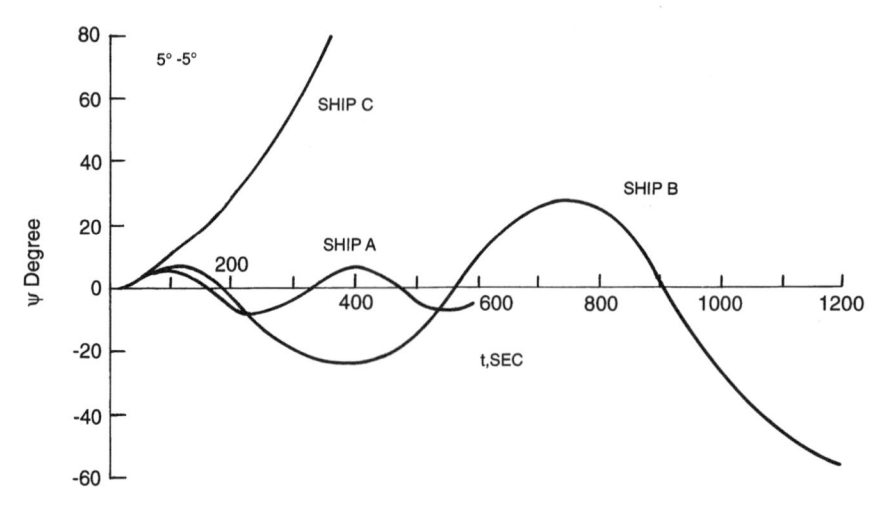

FIGURE D-4 Zig-zag maneuver response: 5–5 degree. Source: Eda and Landsburg (1983).

although scaling and viscous effects limit trajectory predictor accuracy. Mathematical models must be carefully adjusted for variations, and usually separate model tests are required to obtain proper coefficients for use at different drafts or trims of the ship to provide accurate modeling. Properly modeling the level of dynamic course stability is very important in developing a model that will provide the proper vessel response to ship control actions.

Stopping, Backing, and Operating at Slow Speeds

The quickness and distance involved in stopping a ship is related primarily to its mass, astern power, machinery plant, and gearing. Steam turbines typically provide only 40 percent of their ahead thrust capability while astern. Diesel propulsion plants deliver about 80 percent of ahead power astern. The number of engine starts possible from the compressed air system is important for the direct-drive diesel. Once the air supply is expended, quick reversal of the engine is not possible.

Operation at acceptably slow speeds is also a function of the machinery plant. Direct-drive, low-speed diesels have minimum speeds. The trajectory during a stopping maneuver is very difficult to predict because of the complex flows associated with propeller reversal. Other aspects of slow-speed maneuvering that need to be realistically simulated include the use of tugs, thrusters, and anchors to control or impede the ship's movement.

MODELING OF SHIP DYNAMICS FOR SIMULATION

The basic capability that makes simulation possible is that of modeling (predicting and imitating) the trajectory of the ship under various conditions of the environment and controlling actions and forces. This section discusses the present state of ship trajectory modeling. It also addresses the modeling process for ship dynamics and examines issues of accuracy of trajectory prediction and the difficulty in the use of these models for training.

Accurate modeling of ship maneuvering behavior requires an understanding of and a predictive capability for the significant physical forces involved. This understanding may be based on empirical knowledge gained from full-scale and model experiments or on theoretical descriptions of the pertinent physical processes. In practice, both approaches are useful and are often employed in complementary roles.

Physical Modeling

Technical Factors

The principal mechanisms affecting the fluid pressure acting on a ship, and thus its dynamics, are wave effects and fluid viscosity (friction). To correctly represent wave effects, model tests are performed with Froude scaling (model velocity reduced in relation to the actual ship by the square root of the ratio of the model and the ship's length). It is impossible to correctly scale viscous forces, which are exaggerated at the model scale.

Scale models of ships are used in towing tanks to make engineering estimates of various hydrodynamic parameters, including the resistance (drag force) in calm water and motions in waves. Free-running models with self-propulsion and radio control are used to assess standard maneuvering characteristics, such as turns and zig zags. Captive models are more useful in developing models because different components of the forces and moments acting on the ship can be measured separately, during prescribed maneuvers, and used in more general contexts to reproduce arbitrary trajectories, as is required in simulation.

Application of Physical Models to Training

Sufficiently large-scale models with self-propulsion and steering can be maneuvered by an onboard crew to simulate actual vessel maneuvers in suitable lakes or other sheltered waters. Manned models offer the advantages of relatively accurate hydrodynamic representation and realistic scenarios, particularly for low-speed operations, berthing, and ship interaction forces.

The hydrodynamic representation of vessel maneuvering behavior is reasonable because actual hull forms are used at the appropriate scale with appropriate propulsion and control systems. The model, because it is in a real, albeit scaled, operating environment, allows the trainee to physically and visually sense all forces acting on the vessel. For example, the trainee can observe water movement alongside the vessel, propeller wash, and other cues that are normal to ship operations. Manned models also provide a useful platform for special evolutions, such as anchoring, "dredging" the anchor, and "backing and filling" maneuvers (also referred to as "standing round turn"). Because the strength of scaled anchors and ground tackle is not sufficient for the forces that are applied, the anchors are not in scale. Therefore, anchors and ground tackle heavier than scale are used. This practice, however, does facilitate the realistic "hands-on" anchoring practice, which at this time is not mathematically well modeled.

Events take place more quickly aboard a scale model than aboard a real ship because of the scaling factors. Scaling results in roughly a 5 to 1 compression of time (depending on the model scale). The compression of time forces the trainee to interpret and respond to the operational scenario more quickly than in real life and significantly shortens the time required to perform the exercise.

Some advocates of computer-based ship-bridge simulators and mathematical modelers tend to view the need for accelerated interpretation and response as detracting from the training value of the simulation because the trainee is forced to react in an unnatural way. Operators of manned-model facilities believe that the accelerated time frame actually enhances performance, because once back on board ship, mariners can think through the situation more quickly and have more time than in the training environment to observe the operational situation develop and to refine their response. Regardless, experienced mariners who have participated in manned-model training generally have found it to be a useful, limited-task, shiphandling training device, particularly for understanding basic and complex maneuvers, with benefits to real-life performance.

Disadvantages with manned models can include the exaggerated effects of wind and practical limitations in providing different ship hulls, channel configurations, bank effects, and currents. A manned-model training facility is limited to the ship models that are in its fleet (from a low of four to a high of nine). New models can be built at an initial cost of approximately $100,000 each, depending on the sophistication of the model.

Only one of the three operating manned-model facilities is able to control water levels, and thus underkeel clearances. The ability to simulate narrow channels, bank suction, and squat effects is also constrained, not by the models, but by the physical layout of the body of water used for training. One facility has a canal with controlled water depths. Currents are sometimes generated in the canal by an outboard engine mounted at one end. Another facility simulates a canal by anchoring sunken pontoons in a lake, but lacks the ability to control water depths, and thus underkeel clearances. The third facility is constructing an artificial canal.

The improper scaling of viscous forces has not been found to be a serious impairment. Scaling factors, however, do introduce inaccuracies in maneuvers and may be particularly important when involving very small underkeel clearances.

The compression of time can be detrimental for trainees who are either unable to adjust to the scaling factors and relate them back to actual vessel operations or who lack the depth of operational experience and frame of reference that would facilitate the necessary adjustment. Nevertheless, instructors at the manned-model facilities report that time compression has not been a problem during training, nor have they received reports of problems in the real world. Trainees who typically use these facilities, however, are experienced mariners and pilots.

Interactive Effects Using Manned Models

Ship-to-ship interactions can be represented by pairs of manned models operating in proximity. The relevant hydrodynamic interactions between the ships apply if the models are maneuvered in the same manner as the full-scale vessels and if complex trajectories, such as the Texas maneuver, can be simulated. The accuracy of the modeling is affected by viscous scaling effects. The level of accuracy relative to the full-scale vessel has not been formally established. Accuracy levels, however, do not seem to adversely affect training, at least in those cases where the trainee has a full-scale frame of reference for interpreting and correlating the results of manned model to the real world.[4]

Interactions in shallow or restricted water require the correct bathymetry to be reproduced in the simulator facility. This reproduction has only been done in a few cases with idealized representations of confined waterways. The principal and practical limitations occur in a facility's ability to provide or construct the essential physical operating environment and to control the water levels at their facilities, and thus the underkeel clearances.

Mathematical Modeling

If hydrodynamic and aerodynamic forces and moments exerted on the ship can be reliably estimated, the ship's dynamic response can be calculated by integrating the equations of motion, a set of coupled differential equations, at successive time steps. These equations of motion simply state that the product of

---

[4]Except for the U.S. Navy's closed manned-model facility, the majority of trainees have been senior mariners—marine pilots, master mariners, and chief mates—because of perceived training needs by parent organizations, the limited training capacity of manned-model facilities, and costs. Individuals with limited experience rarely have opportunities to attend manned-model courses. The staff officer for this study visited the Navy's manned-model facility and one commercial facility. Although having had prior seagoing service aboard small naval ships, this experience was neither sufficient nor recent enough to establish a frame of reference relative to the maneuvering behavior of large commercial ships. Instead, the staff officer related directly to the maneuvering behavior of the model rather than real-world conditions and did not find it necessary to adjust to the reduced scale. The safety implications of manned-model training for individuals who do not have a well-established frame of reference prior to attending a manned-model course are not known.

the ship's mass and acceleration must be equal to the sum of all relevant forces and moments. If the mass and forces are known, the acceleration can be computed and integrated in time to give the ship's velocity and position at a sequence of time steps that are sufficiently small to approximate continuous motion. Such calculations are easily performed on contemporary digital computers.

The principal problem is in the prediction of hydrodynamic effects (i.e., the force and moment exerted on a ship hull as it moves in a time-varying manner through the water). The accepted practice is to perform tests in a towing tank with models of representative ships performing specified motions, measure the resultant force and moment, and generalize these to other motions using empirical formulas.

General Forms of Mathematical Models

All mathematical models are developed based on Newton's second law that the force acting on a ship is equal to the product of its mass and acceleration. This law applies to all three components of motion—longitudinal, transverse, and vertical (surge, sway, and heave). Analogous formulas apply to the moments and angular motions about each of the three corresponding axes (roll, pitch, and yaw). The equations can be written for the 6 degrees of freedom; however, heave and pitch are usually not included in maneuvering simulations, and roll is only sometimes included.

Hydrodynamic forces from the ship's hull, propeller, and rudder are added to the pertinent external forces, which are due to the effects of factors such as tugs, wind, anchors, and fenders. These forces are related to the dynamics of the motion and the hydrodynamic forces induced in the water.

Various techniques have been used to represent these forces and relationships, with various levels of physical and mathematical complexity. The linear forces, which are proportional to the ship's velocity and acceleration, are relatively well understood. Nonlinear effects involving products of these quantities are more complicated, and ultimately are of practical importance in many cases. Early work in this field represented all of these hydrodynamic forces by polynomials with constant coefficients, determined from judicious combinations of theory and experiment for each ship to be considered. Newer approaches, such as those used at the Danish Maritime Institute, use a look-up table and modular data tables so that empirical data can be modified more easily. The trend toward modularization and more physically based models is discussed later.

Determination of Hydrodynamic Coefficients

Much progress has been made in developing formulas to predict the hydrodynamic coefficients for the linear inertial terms of models. The other coefficients, particularly those that are nonlinear, remain difficult to predict. Tests in which the model is moved in particular motions are used to measure forces on the model. The particular motions eliminate various terms in the mathematical model and allow determination of the coefficients for the active terms.

Some coefficients are developed by towing the model along a straight path at different hull-drift and rudder angles. Special turning basins may be used to obtain similar data when the vessel is in a steady turn. The most comprehensive test procedure is to use a planar motion mechanism. Use of this mechanism permits the model to be driven in oscillatory motions to obtain relatively complete sets of hydrodynamic coefficients both for steady turns and for general unsteady maneuvers.

Most experiments are conducted in wide, deep towing tanks to simulate maneuvering at sea. Insufficient full-scale test data are available for shallow- and restricted-water situations to validate modeling accuracy throughout these regimes. This lack of knowledge is matched by limitations in knowledge about the waterway geometry and current data that are available. Therefore, high confidence is not possible in the trajectory accuracy replicated by the mathematical model in narrow channels and shallow waters with small underkeel clearances.

The accuracy of simulations based on this approach depends on validity at a range of operating conditions, which may differ significantly from the experiments. Obvious examples are when a ship is maneuvering in a channel of restricted depth and width. In such cases the usual practice is to multiply by correction factors or compute forces and moments based on empirically derived functions. This approach accounts successively for the separate effects of the bottom and individual channel sides, notwithstanding their complex interrelationship, and the topology of realistic channels having nonuniform banks. It is here that the hydrodynamic forces pertinent to simulations are most uncertain.

Propeller, Rudder, and Other Control Forces

The correct dynamic modeling of propeller, rudder, and other control forces—such as bow thrusters, and their interrelationships with each other and with the hull—is another difficult task. Generally, modelers have used forces of propellers developed in four quadrants (ship moving forward, propeller thrust aft; ship moving forward, propeller thrust forward; ship moving aft, propeller thrust forward; and ship moving aft, propeller thrust aft). It is assumed that these steady-state forces can be added to other forces acting on the hull. In dynamic situations in which the propeller and rudder are being pressed to elicit particular ship movements, however, unsteady dynamic effects may be important. Uncertainties exist regarding the level of accuracy provided in these areas.

Thus, while there is a wealth of data on the performance characteristics of various propellers and rudders in free-stream conditions, the ability to use the data to build up a maneuvering trajectory prediction with the proper propeller, rudder, and hull interactions has not been developed. More effort must be placed in developing a series of hull-form tests and in determining the interactions among the hull, rudder, and propeller so that the effect of these elements can be determined without retesting. Alternately, use of advanced computational methods can be applied in the near future to develop interaction theories and hull-series relationships.

Bow thrusters are important for slow-speed maneuvers, but their effectiveness diminishes with an increased ship velocity. It is difficult to precisely account for this loss of effectiveness. The use of tugs to assist in ship maneuvers involves complex hydrodynamic and line-force interactions, with substantial changes in the effectiveness of the tugs, depending on their orientation and thrust direction relative to the ship. Anchors can also be used to assist in stopping ships at low velocities, with the effectiveness depending on the characteristics of the bottom, the weight and type of anchor, length and weight of the chain, and the ship's speed over the bottom.

Shallow Water, Restricted Water, and Ship-to-Ship Interactions

The modeling process becomes even more complicated when the effects of shallow water, restricted water (including channel side and bank effects), and ship-to-ship interactions are included. A number of model tests have been performed in shallow water and empirical and theoretical relationships developed. Extensive, full-scale trials of the *ESSO Osaka* provide data for correlation and verification of maneuvering-track accuracy with respect to unrestricted shallow water (Crane, 1979; Crane et al., 1989). The trials were run at deep water and at two relatively shallow drafts. From these trials, it was found that the effects at very shallow underkeel drafts are substantial.

One important restriction is that the underkeel clearances in these trials, and in most towing-tank tests, are limited by concerns about grounding. Ships operate routinely with smaller underkeel clearances than have been reproduced in these trials and experiments. The effect of small bottom clearances requires further full-scale efforts and experiments to understand properly (NRC, 1992).

Restricted-water situations are even more difficult to analyze. The simple case is a canal where the water depth is constant and the sides vertical. When sloping and nonuniform sides are involved, or simply randomly oriented banks, the situation is more complex. Tests performed with various bank configurations have resulted in a procedure for analyzing bank effects (Norrbin, 1974). Extensive model tests have also been performed specifically for the Panama Canal Commission and the U.S. Army Corps of Engineers.

Ship-to-ship interactions also are complex. There are suction forces and repelling forces that take effect and vary as the ships pass each other in narrow channels. The so-called Texas maneuver is an extreme maneuver regularly practiced in the Houston Ship Channel, which is quite narrow. In this maneuver, two ships intending to pass each other steer directly toward each other. As they close on each other, they each initiate a relatively small turn, relying on hydrodynamic repelling forces to push their bows apart. As they pass, suction forces (bow to stern) "steer" the vessels back toward the centerline. Similar suction and repelling forces act on the ships from the proximity of the banks to assist with the maneuver. Analytical methods to treat such phenomena exist, but refinements are needed to account for the many nuances related to the particular distances between objects and the detailed flow characteristics, which can change considerably depending on the channel characteristics.

Model tests to study ship interactions, which are generally performed with two ships on parallel trajectories at a fixed separation distance, do not adequately represent the more-complex trajectories involved in the Texas maneuver. Thus, when simulators have been used to instruct shiphandlers in this maneuver, empirical modifications have been made in the model to achieve more-realistic results.

Modeling of Environmental Influences

Because environmental influences, such as wind, current, and waves, affect the maneuvering of ships, it is necessary to represent the effects of these influences realistically in simulations. When the magnitude and direction of the resulting forces are known, incorporating these physical forces into the simulation is straightforward. Uncertainties exist when the environmental effects are variable, for example, when a ship is passing through an area where there are rapid changes in the strength of the wind or current due to land features or bathymetry. The nonuniformity of the current must be properly represented in confined waterways to obtain realistic operational situations.

Physical models have more difficulty in replicating environmental influences than do mathematical approaches. Generally, there is no systematic effort to represent these effects in manned-model training, although currents and waves can be generated, to a limited extent, by one facility. As a practical matter, manned-model facilities must adjust to existing natural conditions. Facility operators assert that varying conditions are characteristic of and consistent with actual operations to which mariners must routinely adjust. Therefore, varying conditions do not adversely affect training, as long as conditions are not so extreme as to create unrealistic scaling effects.

Mathematical models of water currents can be developed using either appropriately developed two- or three-dimensional numerical models of rivers, bays, estuaries, and coastal areas. These flow fields can then be interpolated to determine current variations across and along a navigation channel and incorporated into a simulation model of the navigation environment. The better simulator ship models use these velocity patterns to compute a relative ship velocity to water, which is the basis of the hull, propeller, and rudder-force computations, and to compute additional drag or drift forces and moments on the ship to account for the moving current.

Because wind effects are exaggerated on small physical models, it is common to assume that this exaggeration of conditions provides a conservative approach for training. Physical models are occasionally used when designing waterways.

VALIDITY AND VALIDATION OF MODELS

Accuracy Requirements

Consideration of model accuracy requirements and validation for ship-bridge simulator training is a subject that generates widely differing opinions. The level of model accuracy required, and the extent to which accuracy needs to be validated, depends on what training or testing is performed using the simulator. Generally, mariner instructors believe that cadet, rules of the road, and bridge resource management training do not require high levels of accuracy, only behavior that is qualitatively correct. For teaching basic shiphandling in deepwater operating conditions, moderately accurate ship hydrodynamic models may be adequate.

High accuracy it generally required for marine pilots and experienced deck officers, who perform at a much higher level of detail and precision in confined waterways than do deck officers generally. Thus, in the context of training for pilotage and specific ship and port operations, the ship, channel, and environmental models need to be as accurate as possible. Pilots will reject simulations if they are not thought to be accurate or true to life. Anecdotal information from operators at simulator facilities suggests that pilots' judgments in this respect may be imprecise or vary among individuals because of different maneuvering strategies for the vessel or waterway being modeled or levels of experience. Furthermore, a particular vessel's maneuvering capabilities may not be fully appreciated; mariners do not routinely operate at the extreme limits of vessel maneuvering capabilities, although the nature of piloting typically brings marine pilots much closer to these limits. For these reasons, a simulation that is modified to meet the expectations of a single expert may not be representative of adjustments needed in the numerical model.

If marine simulation is to be used for the licensing and qualification of shiphandlers, the performance of the ship in the environment in which the operations are to be modeled would benefit from being highly accurate with respect to trajectory prediction and to overall realism needs. Trajectory prediction accuracy in the simulation of vessel maneuverability could avoid creating unrealistic operating conditions that could influence an individual's performance. On the other hand, an individual's familiarity with vessel maneuverability in a particular waterway could potentially mask weaknesses in shiphandling knowledge, skills, and abilities.

Model Validation

Model Adjustments by Simulator Facilities

A properly developed mathematical model will produce, with acceptable accuracy, the motion of a vessel in response to physical forces and maneuvering commands. The validity of the model in various modes of ship operations needs to be established before the model is applied in training. Once a model's validity has been established, it would not be appropriate to modify the model unless some discrepancy in performance were subsequently disclosed. In the latter case, adjustments need to be made by qualified individuals. In every case, model tuning needs to be revalidated for reasonableness and physical and hydrodynamic correctness. Revalidation in those cases in which simulation was officially used in marine licensing would need to meet criteria established for this purpose.

Some guidelines exist for adjusting ship models for different drafts or trim or for adjusting size for geosims (i.e., a ship hull that is geometrically identical but different in size). Some of these guidelines are documented; others are not. Models based on physical tests (such as hydrodynamic experiments in towing tanks) are considered more robust, particularly with respect to modifications. Models based primarily on mathematical curve fitting of test data from different models require careful attention to the development and modification of their coefficients. In each case, validating the models with respect to their ability to produce realistic maneuvering behavior is subjective, because a scientifically based validation methodology has yet to be developed (NRC, 1992).

Ship-bridge simulation facilities that conduct both channel design and training simulations generally, if informally, follow the interdisciplinary validation approach (NRC, 1992). There is, however, a tendency among some simulation facilities to shortcut the validation process. Instead of using an interdisciplinary team, modifications by the instructional staff are permitted, especially for models that have been generated by the staff using modeling routines available in some of the ship-bridge software packages. Facility-generated models of this type are usually developed because a needed model is not available in the facility's software library or because the available model did not perform to the satisfaction of the trainees or instructional staff. The result is that individuals who are self-taught modelers sometimes create and then validate the models that they have developed without the benefit of interdisciplinary perspective to ensure the overall reasonableness and accuracy of the simulated vessel's maneuvering behavior. The involvement of a large number of mariners in validating the accuracy of trajectory prediction may be useful, but such a practice should not be a substitute for including hydrodynamic expertise in the validation process.

There are strong hydrodynamic reasons to discourage field adjustments of mathematical models and ad hoc creation of mathematical models using resident software. Occasionally, however, ship-bridge simulation facility operators are motivated to take such actions and through trial and error sometimes achieve favorable results with respect to the realism of vessel maneuvering behavior.

Motivations for field modeling might include anecdotal indications of lower fidelity in maneuvering behavior than needed to achieve training objectives, lack of support or untimely support from simulator manufacturers for model corrections, and lack of resources needed to develop a model to meet client application needs. Field adjustments are not good practice from a hydrodynamics perspective unless undertaken by an interdisciplinary team of individuals qualified in both the operational and hydrodynamic elements of the simulation or the field adjustments are subsequently reviewed by a suitably qualified hydrodynamicist to validate hydrodynamic fidelity. This approach would also be appropriate for deliberate departures from realism that include adjustments to the mathematical model.

Adoption of interdisciplinary validation of adjustments to the trajectory prediction model would help ensure that marine simulation operators have as complete an understanding of the capabilities and limitations of their simulators and simulations. This understanding needs to be appropriately conveyed to trainees to minimize the potential for training-induced error. The capabilities and limitations of a simulator, and the resulting simulation, are also important considerations when using marine simulation to meet marine licensing requirements or applying it directly within the licensing process—for example, for competency assessments.

It should be noted that it is not only the ship model that should undergo validation, but also its environmental segment. The model of currents, channel definition, wind, and wave conditions can be independently verified; but the entire simulation scenario model should also be validated.

Correcting Inaccuracies in the Mathematical Model

When ship performance is found or suspected to be incorrect, the base cause needs to be identified. A series of tests can be performed to isolate deficiencies that need to be corrected. Whether the shortcoming is a ship characteristic, something in the channel or operating environment description, or a combination needs to be determined.

Obviously incorrect simulator behavior can normally be detected by an experienced mariner; however, the cause of the problem may not be readily understood by the mariner. More-subtle problems normally require detection by an individual with expertise with the hydrodynamics and simulated vessel, and, in the case of specific waterways, the route. Normally, this detection can be accomplished using marine pilots from the area being simulated. Occasionally, marine pilots may not be familiar with the vessel being simulated. In these cases, master mariners can be involved for the necessary expertise. The mariner needs to supply an accurate description of the correct and incorrect behavior. Because individual perceptions, expectations, and maneuvering strategies vary, it would be

APPENDIX D   233

good practice for another mariner or mariners to also describe the problem to verify its existence.

If the problem pertains to the characteristics of the ship, adjustments to the model are most appropriately performed by individuals familiar with it, because the adjustments in any coefficient need to be correlated with their effects on other coefficients. The knowledge required to adjust the coefficients necessitates the involvement of a hydrodynamicist, as well as a mariner expert in vessel behavior. If the problem is with the channel or environment, the adjustments would require expertise in the modeling of channel design effects on vessel maneuverability.

The judgment of an experienced pilot or mariner may not be correct if he or she is not familiar with the specific ship type, size, and hull form. Vessel size and minor changes to the hull afterbody can result in quite different vessel maneuverability characteristics. Comparative model tests may be required to verify expected vessel maneuverability.

FUTURE DEVELOPMENTS

Vessel maneuvering prediction modeling is a developed science that provides highly useful tools for building marine training simulators. From a technology perspective, the future is promising for improving the accuracy, flexibility, and extent of simulator-based training applications. Theoretical and numerical methods are powerful and nearing practical application. Computational power, graphics, and multimedia capabilities and the proliferation of powerful microcomputers enable the use of sophisticated mathematical trajectory models and new approaches for training applications.

Computational Ship Hydrodynamics

Theoretical and Numerical Methods

It is possible to use theoretical and numerical methods to predict hydrodynamic effects pertinent to maneuvers. This procedure depends on contemporary developments in the field of numerical ship hydrodynamics aimed at replacing estimation and synthesis based on towing tank tests. Advantages of the numerical procedure include the ability to model both the ship and waterway bathymetry in their actual configurations. Ship-to-ship interactions can also be treated in their precise time-varying relationship, overcoming the limitations of straight parallel courses.

Viscous Effects

In the absence of significant viscous effects, potential theory can be used. Substantial progress in this area has been made using panel methods where the ship's surface (and surrounding channel topography in some cases) is represented by a large number of small panels. Such methods give useful predictions of ship interactions in channels (Kaplan and Sankaranararyanan, 1991).

When viscous effects are important, as in cases of very small underkeel clearance or low-speed maneuvering, it is possible to use three-dimensional solutions of the complete equations of motion for the fluid, including viscosity (the Navier–Stokes equations). Developments in this field are currently expanding (see NRC, 1994). The computational burden is substantial, and implementation to real-time simulations will require future improvements in computer hardware. Navier–Stokes solvers may be useful in special applications—for example, to study the effects of very small underkeel clearances where viscous effects are important and where the computational domain is relatively compact. These applications could then lead to improved functional relations that could be used in modular real-time simulator models.

Some "zonal" computations have been performed, which combine a viscous solution close to the ship with a simplified solution far away. This approach has been used for research programs dealing with steady motions of ships to predict the drag force (Larsson et al., 1990). Extensions to unsteady yawed motions are anticipated and may offer an effective compromise between the relative simplicity of potential theory and the computational demands of Navier–Stokes codes. While these methods are as yet unproven in the context of ship maneuvering, they have been extensively developed for research on steady-ship resistance, ship motions in waves, and propeller hydrodynamics. It is possible to include all these effects in a comprehensive model for maneuvering simulation.

The effect of waves on ship maneuvering is considered by Ottosson and Bystrom (1991). Special examples in which wave effects are important include training of shiphandlers for extreme storm conditions (encountered rarely but where experience in safe headings and maneuvers may be vital to the safety of the vessel) and the navigation of ships in entrance channels of restricted depth when waves are present.

Microcomputer-Based Shipboard Simulation

Although physical models, shiphandling simulators, and radar simulators will continue to be the major types of simulation available for full-mission, multi-task, and limited-task training, new approaches will expand the traditional media that are available for training. The desktop microcomputer with standard PC compatibility has achieved sufficient computing power to offer a platform for real-time (and fast-time) simulation. PC-based training software suitable for use on board or ashore have already been developed. Onboard PC simulator training capability can also potentially complement the use of trajectory prediction capabilities from such simulations for real-time navigation of the ship. Such simulation applications will lead to the need for high-fidelity trajectory prediction models that are ship-specific and which can adjust to restricted waters with bank, shallow-water, and ship-to-ship interactions.

Microcomputers are becoming more common on board ship, so that it is technologically possible to perform simulations for training, practicing, and previewing specific transits.

The modeling of ship dynamics, including channel bathymetry and other ships, can be accommodated within the computational resources of contemporary desktop microcomputers. Principal limitations are with user interfaces, including the bridge mockup, controls, and visual images. The user is placed in a simplified training environment, in contrast to a situation in which the participant is placed inside a more complete training environment associated with a full-mission ship-bridge simulator. The same mathematical model can be used to drive virtually any form of computer-based marine simulation.

Future trends are expected to enhance microcomputer simulation capability. It is already possible to interface shipboard computers with modern ship systems and instrumentation. The actual bridge controls and instrumentation could be used (during periods when it would not conflict with normal navigation) to provide a realistic operating environment, except for the visual image. The use of this equipment by on-watch personnel could, however, be viewed as distracting them from their primary watch responsibilities. It is also possible to derive ship-specific inputs (normally derived from towing-tank tests) directly from operational measurements. For example, correlations of the ship's velocity, propeller revolutions, and rudder angle with accurate measurements of position and heading offer the possibility to derive hydrodynamic maneuvering coefficients that could be more accurate and specific to the ship and its environment. Onboard acquisition of such information can provide improved modeling of specific ship and waterway conditions. Previewing ship movements would then be even more useful for training. Such extensions of existing methods could also be used to enhance the fidelity of land-based simulators.

Physical Versus Mathematical Models

Physical models provide a simple approach to imparting knowledge on the hydrodynamics of ship motions in deep water and close-in operations, including docking, coming alongside, shallow water, banks, and ship-to-ship operations. Action is quick, several times faster than real time; for example, it is 5 to 1 for a model at 1 to 25 scale. The compressed time frame and layout of the training facilities result in virtually continuous training of the mariner while operating the manned models. The scaling of the dynamics results in propulsion and rudder-action modeling inaccuracies, and human stereoscopic vision is more acute than on the ship because of the spread of human eyes when related to the scale of the model. Currents, wind, and waves are difficult to model, and usually done to only a limited extent, and then not at all facilities.

Computer-based simulations have considerable flexibility in that a situation can be replayed or played with changes in situation or reality. Mathematical models are more accurate and reliable in deep-water situations. Their abilities in shallow water and bank situations are more uncertain, and there is insufficient full-scale data and proper bases for accurate prediction in these conditions. Significant progress has been made in this area.

The ability to model rivers, estuaries, and coastal areas is continuing to improve with the continued development of three-dimensional models that can more accurately reproduce complex bathymetry and current patterns. Validation is a problem in that the opinion of pilots or experts on how the ship reacts in the real world are used to make adjustments to the model. These adjustments may or may not be correct. How the adjustments are made may vary from facility to facility. Standards need to be developed and guidance provided to simulator users so that inappropriate adjustments are not made. Further full-scale measurements of ship maneuvering characteristics are needed to validate the theoretical and mathematical models and to provide a model that is defendable from expert criticism.

The rapidly expanding availability of differential global positioning system reference stations and systems provides great promise for obtaining accurate measures of ship behavior in restricted waterways that have heretofore been difficult and not economical to obtain.

Validity Requirements

From the modeler's perspective, the simulator user must specify what accuracy is needed for particular training objectives. The simulation modeler, physical or mathematical, can then assess whether that accuracy can be provided. Pilots, for instance, need very accurate models to properly portray bank and other sophisticated situations, whereas a less robust model may suffice to introduce very basic operational concepts and procedures to beginners. Regardless, model validation requirements need to be based on the training objectives that the simulation is intended to support.

The use of an expert to validate a model is also problematic when the ship differs from the one the expert is familiar with. A similar appearing vessel can have significantly different maneuvering characteristics because of different levels of dynamic stability due to design of the hull or appendages. A change in vessel size also can provide quite different responses.

The ship-bridge simulator appears to be able do everything. It has all of the controls and appears to the trainee to be real. The trajectories provided by the model may be highly accurate for many situations, but could be totally wrong in others. Experts in vessel operations could provide the modelers with knowledge of where their models fall short and warn trainers of conditions where the model's behavior is sufficiently different from actual behavior of a real ship.

Operational Environments

There has been a substantial development of simulated scenario areas for U.S. ports and waterways accomplished by the U.S. Army Corps of Engineers for navigation project improvements. Navigation environments that may have been modeled for other purposes could also be used or modified to develop navigation training scenarios. Consideration should be given to how these models could be used in developing verified environments for training requirements.

Flexibility Needs

Numerous mathematical models are currently used to drive various simulators. Hydrodynamic coefficients used in the models cannot be easily exchanged. If one simulation facility has a need for a ship's model it does not have, acquiring it is difficult. A standard method for exchange of models or modeling coefficients would facilitate the sharing of important technical information. Modularity of the models is also needed so that a module can be replaced with more sophisticated versions when available without requiring overall modifications.

Validated and practical theoretical solutions for coefficients are needed, particularly to represent nonlinear hydrodynamic effects. Such formulations would provide a more objective basis for models and help with validation by supplementing the subjective validation of model (and simulator) performance. The more physically based models can be, the more flexible and readily adaptable the models will become.

Software Issues

As with any large computer program, the software developed for ship simulation is subject to various potential limitations. These include:

- errors or restrictions in the assumed hydrodynamic model,
- numerical errors due to the reduction of the model to computational form (such as time steps that are too large), and
- programming errors (bugs).

As programs become older, age and insufficient maintenance may inhibit their relative quality and relevance, unless special efforts are invested to provide updated versions. Computer programs developed for ship simulation require substantial investments of expertise and effort, and this investment is normally protected by licensing agreements. Further protection is often achieved by distributing the code in an executable form that cannot be modified or transferred among different computational environments. Public-domain software is a preferable alternative from some users' standpoints and offers the significant advantage that it can be shared within the simulation and hydrodynamics communities to enhance exchange of ideas, correction of errors, and improvement of hydrodynamic models. This concept of "open" software is particularly suitable if simulation is associated with mandated requirements and, more generally, to facilitate simulation validation.

Research Needs for Model Improvement

Computational fluid dynamics holds great promise as the solution to current modeling problems. Practical developments and use in bridge simulators cannot be implemented until existing research codes are applied to the specific computational tasks of maneuvering simulation. Two different levels of implementation can be envisaged.

At the first level, computational methods can be used to complement towing-tank testing in the determination of hydrodynamic coefficients. Examples where computational methods can play a useful role include ship-to-ship interactions in restricted channels and the use of Navier–Stokes solvers to consider ship operations with small underkeel clearance in shallow water.

At the second level, where computations of pertinent hydrodynamic effects can be performed with sufficient accuracy and speed, this approach can replace the traditional approach based on curve fitting. This may be particularly useful for ship interactions and restricted water effects, where existing simulator models are severely limited. The possibility now exists, based on three-dimensional panel methods, to compute the relevant interaction forces and moments during the simulation, based on the actual ship trajectories and channel topographies, and avoiding the uncertainties and limitations emphasized above.

Current mathematical models would benefit from being modularized and validated in parts so that the science of modeling can progress and accuracies can be better established. Full-scale experiments also are needed to advance the state of practice in modeling, particularly for shallow water and restricted waters with banks. The process of certification of models and validating changes needs improvement. Modeling of operations at slow speed and with reversing propeller situations also needs improvement.

With respect to mariner licensing, the lack of standardized mathematical models would make it difficult to ensure the equivalency of the vessel maneuverability basis for evaluating mariner performance. To the degree that vessel maneuverability could affect individual performance during evaluation, it would be beneficial to provide standardized ship models. A standard set of harbor operating conditions could be developed for specific ships that can be accurately validated against performance measurements and used as a consistent basis for assessing mariner competency. Since accurate ship models and channel environment models are required for some purposes, it would seem reasonable to use such models for all purposes.

MODELING REQUIREMENTS FOR SIMULATOR-BASED TRAINING

The appropriate modeling approaches, their levels of accuracy and sophistication, and their utility for accomplishing effective simulation training conceptually depend on the specific training objectives that need to be satisfied. As a practical matter, simulation capabilities are developed independently of training objectives. As a result, it becomes necessary to compare training objectives to available simulation resources to determine the suitability of simulation as a training medium or of specific simulation resources to meet specific training objectives.

Manned models offer an alternative to on-the-job training, subject to the trainees ability to adjust to the scaling factors and then to correctly translate the lessons learned back to the real world. As suggested earlier in this appendix, the manned models are highly suitable and effective in instructing the experienced mariner and pilot about the principles of ship maneuvering and hydrodynamic interactions. Simulation-based shiphandling training for specific ports and waterways must necessarily rely on ship-bridge simulation capabilities with mathematical models. In this case, it is important that the accuracy of the trajectory prediction model is understood by all involved in a simulation so that false expectations are not created relative to real-world operating conditions.

The modeling of towing vessels alongside or pushing ahead and of integrated tug-and-barge combinations has not received the attention that has been given to the modeling of ship dynamics. Towing-vessel control modules with flanking rudder capabilities are available at only a few marine simulation facilities, limiting the availability of simulation training for towboats that routinely use this equipment. The type of operational situations that these vessels function in are almost always restricted waterways with strong currents, thus requiring highly developed environmental models for accurate training situations.

•

REFERENCES

Crane, C.L. 1979. Maneuvering trials of a 278,000 dwt tanker in shallow and deep water. Transactions of the Society of Naval Architects and Marine Engineers 87:251–283.

Crane, C.L., H. Eda, and A.C. Landsburg. 1989. Controllability in Principles of Naval Architecture. Jersey City, N.J.: Society of Naval Architects and Marine Engineers.

Eda, H., and A.C. Landsburg. 1983. Maneuvering performance analysis during preliminary ship design. Pp. 179–186 in Proceedings of Second International Symposium on Practical Design in Shipbuilding. Tokyo: Society of Naval Architects of Japan and Korea.

Graff, J. 1988. Training of maritime pilots—the Port Revel viewpoint. Pp. 62–76 in Proceedings of Pilot Training, Southampton, England, July 12–13.

Hays, R.T., and M.J. Singer. 1989. Simulation Fidelity in Training System Design: Bridging the Gap Between Reality and Training. New York: Springer-Verlag.

Kaplan, P., and K. Sankaranararyanan. 1991. Theoretical analysis of generalized hydrodynamic interaction forces on ships in shallow channels. Transactions of the Society of Naval Architects and Marine Engineers 99:177–203.

Larsson, L., L. Broberg, K.-J. Kim, and D.-H. Zhang. 1990. A method for resistance and flow prediction in ship design. Transactions of the Society of Naval Architects and Marine Engineers 98:495-535.

Norrbin, N.H. 1974. Bank effects on a ship moving through a short dredged channel. Pp. 71-88 in Proceedings of the Tenth Symposium on Naval Hydrodynamics. Cambridge, Massachusetts, R.D. Cooper and S.W. Dorof, eds. Washington, D.C.: U.S. Government Printing Office.

NRC (National Research Council). 1992. Shiphandling Simulation: Application to Waterway Design. W. Webster, ed. Committee on Shiphandling Simulation, Marine Board. Washington, D.C.: National Academy Press.

NRC (National Research Council). 1994. Proceedings of the Sixth International Conference on Numerical Ship Hydrodynamics, Iowa City, Iowa, August 2-5. V.C. Patel and F. Stem, eds. Washington, D.C.: National Academy Press.

Ottosson, P., and L. Bystrom. 1991. Simulation of the dynamics of a ship maneuvering in waves. Transactions of the Society of Naval Architects and Marine Engineers 99:281-298.

Puglisi, J.J. 1987. History and future developments in the application of marine simulators, tomorrow's challenge. Pp. 5-29 in Proceedings MARSIM '87, Trondheim, Norway, June 22-24. Trondheim, Norway: International Marine Simulator Forum.

SNAME (Society of Naval Architects and Marine Engineers). 1993. Proceedings of the Workshop on the Role of Hydrodynamics and the Hydrodynamicist in Ship Bridge Simulator Training, Jersey City, October 22. Jersey City, N.J.: SNAME.

APPENDIX

E

Outlines of Sample Simulator-Based Training Courses

This appendix includes the outlines of two simulator-based training courses. These courses are conducted by members of the NRC Committee on Ship-Bridge Simulation Training.

BRIDGE WATCHSTANDING SIMULATION TRAINING COURSE FOR CADETS, U.S. MERCHANT MARINE ACADEMY, KINGS POINT, NEW YORK

U.S. Coast Guard License Programs

Deck Officer: training in nautical science as preparation for the third mate's license examination. (Required of cadets majoring in marine transportation and ships' officers).

Engineering Officer: training in marine engineering as preparation for the third assistant engineer's license examination. (Required of all engineering majors). Dual license majors take both license-preparation instruction and license examinations.

Experience with the bridge watchstanding simulator is a necessary part of the license curriculum; therefore, it must be able to meet the U.S. Coast Guard standards for training.

Bridge Watchstanding Simulation Training Course

Aims

1. To enhance the potential third mate's decision-making skills as they apply to traffic and voyage-planning situations.

2. To sharpen the cadet's bridge watchstanding skills to the highest level prior to graduation.

Objectives

1. Understand the maneuvering capability of own ship (a 40,000 deadweight ton tanker) and amounts of rudder required for various maneuvers, including:

- rudder orders,
- rudder response time,
- heading change rate and its time dependency,
- engine speeds and vessel maneuvering,
- engine orders,
- engine response time and limitations (diesel versus steam),
- speed change and its time requirements, and
- effects of environment (wind, current, etc.) on shiphandling.

2. Understand the importance of monitoring and assessing traffic situations and identify collision risks as it applies to:

- a real-world application of the COLREGSs in clear and restricted visibility,
- proper use of VHF communications,
- the need to maintain an efficient lookout and to take visual bearings to determine drift,
- clear and concise reports to the master using relative bearings and drift,
- timely and substantial alterations of course,
- special circumstances and ambiguous rules-of-the-road situations,
- how to maneuver the vessel in extremes.

3. Keep a safe navigation watch in coastal waters and while approaching a pilot station as it pertains to:

- importance of reading and complying with standing and night orders,
- monitoring all bridge equipment and responding to malfunctions,
- proper transfer of the watch and proper log-book entries,
- communications (inter and intra),
- preparation of vessel for port arrivals and departures,
- preparation of passage plan and monitoring progress of vessel in accordance with the plan,
- identifying situations when the master must be called,

APPENDIX E

- teamwork and role of the officer of the watch with master or pilot,
- utilizing all means available for fixes and parallel indexing (i.e., dead reckoning ahead, turn bearings, plotting advance and transfer, combination of visual and radar fixes, proper utilization of ranges, and parallel indexing.

The Voyage: New York to Port International Scenarios

Week	Time on Simulator for Each Watch Team	
1	30 minutes	Bridge and vessel familiarization.
	45 minutes	Vessel at Stapleton—prepare for sea master on bridge; weigh anchor; full away; master departs bridge (day).
2	1 hour	Prepare for arrival and arrive New York (night); Sandy Hook pilot boards; master in cabin; pilot/watch officer relationship and transfer of watch in pilotage waters.
3 and 4	2 hours	At-sea rules-of-the-road scenarios; various steering failures and reduced visibility (day and night).
5	1 hour	Prepare for arrival and arrive Cristobal anchorage (sunrise); begin grading watch teams.
6	1 hour	Prepare for departure Cristobal (day); depart from Cristobal.
7	1 hour	Transit Singapore Straits and transfer of watch (night).
8	1 hour	California coast watch in vessel traffic lane and transfer of watch (day).
9	1 hour	Arrival preparation and arrival Santa Cruz Channel for Port International (day).
10	1 hour	Depart Port International under Santa Cruz VTS (vessel traffic safety) in a mine-swept channel for sea (night).

BRIDGE TEAM MANAGEMENT COURSE, SOUTHAMPTON INSTITUTE OF HIGHER EDUCATION, MARITIME OPERATIONS CENTRE, SOUTHAMPTON, UNITED KINGDOM

The following are the aims and objectives of the Centre's Full-Mission Ship's Bridge Simulator Course followed by the objectives of the scheme of work for the Bridge Team Management Course. This example illustrates development of an exercise scenario from a single, relevant simulator exercise. The exercise is from the later part of the course, when the group of trainees involved will have developed their teamwork and bridge resource management skills as a result of preceding exercises and lectures and discussions.

Full-Mission Ship-Bridge Simulator Course

Aims

1. To provide participants with the opportunity to experience and analyze various navigational scenarios and demonstrate procedures to assist in the safe conduct of the vessel.

2. To exercise procedures that will take into account all relevant regulations, as well as quality management techniques that work toward high standards of excellence and professionalism.

Objectives

The course objectives concentrate on passage-planning techniques and bridge resource management techniques.

1. Passage-Planning Techniques. Participants completing the course will be able to appraise and formulate a detailed passage plan including but not limited to:

- no-go and danger areas,
- margins of safety,
- track to be made good,
- use of transits (ranges) clearing bearings and distances,
- visual cues,
- parallel indexing,
- conspicuous radar targets,
- course alterations and wheelovers,
- speed and timing of progress,
- aborts and contingencies,
- allowance for squats, and
- interpretation and use of maneuvering data.

Participants completing this course will be able to execute such a plan and demonstrate a good understanding of proper bridge resource management.

2. Bridge Resource Management. Participants completing the course will be able to demonstrate their understanding of:

- the need for good teamwork,
- the need for good internal information flow between all team members,

- the importance of developing the skills and confidence of junior members of the team,
- the need for good management of all available resources (i.e., personnel, equipment, and time),
- good external and internal communication procedures,
- the need to identify the development of an error chain and to break such a chain,
- the importance of traffic management, particularly in confined waters,
- the role of the engine-room staff in the team, and
- the risk factor in an anticipated operation and in the development of good risk management procedures.

Exercise

On completion of the exercise, participants will have:

1. Appraised and planned a complex, confined-waters passage including departure from an anchorage, a difficult turn, and traffic management.
2. Planned for contingencies, emergencies, and an emergency anchorage.
3. Discussed and planned for the integration of the pilot into the bridge team—including correcting any deficiencies in this area highlighted by exercises 4 and 6, as applicable.
4. Conducted an effective ship/pilot exchange of information—in particular, how the ship is to maneuver when leaving the anchorage.
5. Monitored the vessel's position closely when leaving the anchorage and supported the pilot accordingly.
6. Recognized and dealt with various developing traffic situations in accordance with the collision regulations.
7. Maintained the vessel's progress in accordance with the passage plan.
8. Maintained the vessel on the planned track in a narrow channel in strong cross-tides and winds.
9. Responded effectively to an emergency.
10. Brought the vessel to an emergency anchorage.

The master will have briefed his team and will have:

1. Conducted an effective master/pilot exchange.
2. Demonstrated his or her ability to make command decisions.
3. Shown situational awareness and anticipation in dealing with traffic situations and in conducting the navigation in a difficult, narrow channel.
4. Responded decisively to an emergency situation.
5. Delegated navigational tasks effectively.
6. Gained experience in the command role of a bridge team.

APPENDIX

F

Uses of Simulators Illustrative Case Studies

Research results reported in the literature and anecdotal evidence suggest that simulator-based mariner training can be used to improve the development of knowledge, skills, and abilities and that such training transfers to and improves the safety of actual operations. These results have not, however, been conclusively demonstrated through empirical research. Nevertheless, marine simulator-based training is considered of sufficient value by marine educators, some operating companies, and a growing number of marine pilot organizations to motivate their investment in the use of simulators to enhance professional development. The case studies presented in this appendix provide examples of various uses of marine simulators within the marine industry and the piloting profession. The studies provide practical insights on course content and operational scenarios and illustrate the value placed in simulators by the individuals and organizations. Case studies are presented for:

- a cadet watchkeeping course, which illustrates careful attention to instructional systems design;
- the professional development of apprentice marine pilots;
- port familiarization for ship masters and pilots and the development of port-entry protocols;
- the combined use of simulator-based and follow-up, onboard training on coastwise towing vessels; and
- a recently approved combination training and testing course.

CASE STUDY ONE
SHIP-BRIDGE SIMULATOR-BASED CADET WATCHKEEPING COURSE, U.S. MERCHANT MARINE ACADEMY

This case study reports on the curricula for the U.S. Merchant Marine Academy's (USMMA) full-mission, ship-bridge simulator-based watchkeeping course (Meurn, 1990) and its implementation. The recorded grades for 31 courses involving 233 three- to four-person cadet watch teams (approximately 900 cadets) over a 10-year period (October 1985 to March 1994) were reviewed. The objectives are to examine the application of instructional systems design concepts, to quantitatively identify trends in cadet performance over the course of instruction, and to develop insights and lessons about the application of simulator-based training for third mate candidates with limited, prior nautical experience.

The data that were available were not collected as part of a research experiment. There were no control groups, nor was there follow-up monitoring of real-world performance. It was not possible, therefore, to compare course performance data for ship-bridge simulator-based training with the results of traditional training. Also, there were no data to assess transfer effectiveness for the individuals who participated in the training.

The analysis presented in this case study indicates that the development of watchstander knowledge, skills, and abilities can be significantly improved using simulators as a training medium. The data are not available to determine whether ship-bridge simulator-based training is more effective and efficient than traditional training. The analysis does suggest, however, that the ability to control the learning process (including the ability to design scenarios, monitor performance, and debrief cadet participants), in contrast to the lesser control over learning situations aboard ships at sea, leads to improvements in efficiency.

The Evolution of Simulator-Based Watchkeeping Training

The USMMA began using simulator-based watchkeeping training in 1980 to supplement other practical training opportunities. Simulation enabled the cadets to become actively involved in decision making on the bridge, in contrast to the observation status associated with most cadet sea-time training aboard commercial vessels.

The USMMA substituted ship-bridge simulators in form and function for a real ship's bridge. Basic operational scenarios were used. Little attention was given to creating scenario designs that would bring the trainees into specific learning situations with specific training objectives. Instructor-cadet interaction was similar to the relationship found aboard a school ship in which a licensed officer is responsible for safe vessel operation. During drill critiques, the instructor interacted as a lecturer rather than as a mentor or facilitator.

The training program was extensively revised in 1985 to incorporate an innovative instructional design process for watchkeeper training used in marine simulator-based courses at the Southampton Institute in England. The course was and continues to be offered to cadets in their senior year as final preparation for actual watches as third mates upon graduation. The course is given after training assignments aboard commercial ships. Separate departments oversee the watchkeeping training aboard the USMMA's training vessel and the simulator-based watchkeeping course curricula.

Specific objectives were established for each training course, and drill scenarios were carefully crafted to bring the cadet watch teams to the instructor-planned learning experience. The mariner instructors redefined their role as mentors and, during debriefings, as facilitators, to more effectively stimulate trainee intellectual analysis of the results. Each exercise was graded numerically, recorded, and maintained. The instructors reported that as training progressed, they observed a notable difference in the attitude and command (i.e., leadership) presence of the person in charge of the watch. For each task performance, the instructors reported a subjectively measurable increase in the level of confidence. On completion of the training, the instructors observed that professional maturity, attitude, and confidence were markedly improved.

The Simulator-Aided Watchkeeping Course

The simulator-aided watchkeeping course consisted of 10 preparatory classroom lecture sessions, 4 ungraded familiarization scenarios aboard a full-mission ship-bridge simulator at the Computer Aided Operations Research Facility (CAORF) in Kings Point, New York, followed by either 5 or 6 graded watchkeeping drills aboard the CAORF ship-bridge simulator. (The academic calendar and administration of the U.S. Coast Guard's licensing examination for third mate preclude a sixth drill during courses held in the fourth quarter.)

The classroom sessions consisted of case studies, familiarization with standard bridge team and operational procedures prescribed for use by the USMMA Nautical Sciences Division, advance preparation for the drill scenarios, and general briefings and technical discussions.

The simulated ship is the M/V *Capella*. She is a 40,000 deadweight ton tanker with a single propeller and rudder. Her bridge configurations are representative of this type of vessel. Navigation aids include radar, automatic radar plotting aids, Loran, and fathometer. Electronic charting and automated real-time, precision navigation systems are not available.

The familiarization scenarios consisted of a daytime departure from the port of New York, a nighttime arrival at the port of New York and New Jersey, and both daytime and nighttime at-sea scenarios involving rules-of-the-road situations in unrestricted and restricted visibility. Rules-of-the-road scenarios are used because they are usually the first type of situations that third mates experience and must respond to during their first underway watches. The familiarization simulations also focused on how to use the bridge equipment correctly, as well as how to communicate with the captain by the ship's internal communications and by voice radio with other vessels. By the commencement of the first graded drill, the preparatory elements of the course had exposed each watch team to arrival and departure procedures, change of watch, routine watchstanding procedures, and emergency procedures.

The graded simulations consisted of various operational scenarios with specific learning objectives. Each scenario was designed to bring the cadet watch team to the learning situation. Each succeeding simulation increased in difficulty by about 10 percent, as estimated by the mariner instructors who conducted the course. Scenario design took account of the instructor observation that cadets, as new learners, typically were capable of accomplishing a single task well, but experienced difficulty performing multiple tasks concurrently, especially in a stressful operating environment. The scenario designers also considered the fact, known to experienced mariners, that subtle factors that influence operations can quickly combine in their effect to create challenging and even untenable operating conditions. These factors contributed to the decision to use a single ship for all drills and were used as underlying themes in developing scenarios.

Many watchkeeping courses place trainees aboard a variety of ship types during succeeding lessons. Only the M/V *Capella* was used in the USMMA cadet watchkeeping course. The experience of the mariner instructors is that the use of the same platform facilitates cadet learning. The basic training objectives of the course were to build good watchkeeping practices and procedures that could be applied to all bridge situations and to prepare cadets to make timely and correct decisions in a multiple-task environment.

The mariner instructors agreed that changing the ship's operating characteristics for trainees with a limited nautical background would detract from course objectives. Changing the platform would have required the cadets to adjust not only to the changing operational scenarios, but also to changes in the ship and its maneuvering behavior. The use of the same ship contributed to progressive improvements in watch team performance reported in this case study. This approach is consistent with actual operations. Duty aboard the same ship for extended periods and continued service aboard the same category of ship are characteristic of marine operations in general.

Early graded scenarios in the drill sequence provided an opportunity for cadets to become familiar with the simulator environment, with the addition of actual responsibility for performance. Subsequent scenarios increased in complexity and difficulty, progressively combining multiple tasks and maneuvering situations to create mentally demanding and stressful operating conditions. A single scenario, for example, might combine an early morning approach, overtaking another ship, meeting another ship, a maneuvering situation in which a "bail-out" option was not conveniently available, and a watch relief.

Each of the individual elements of the scenario posed no significant problem, and the overall scenario represented circumstances often faced during harbor approaches and departures. The combined effects of the maneuvering situation, however, with the inopportune timing of a watch relief and its distracting aspects, were used to create a challenging situation in which the timing and precision of navigation and conning tasks were critical.

The first three graded drills exposed cadets to basic fundamentals associated with arrival, departure, and watch relief. The final three drills repeated the basics of the first three drills and added additional degrees of complexity, such as crosscurrents and wind effects. The first drill was a straightforward port arrival without current and wind effects. The second drill was a departure from an anchorage for a berth. This scenario exposed cadets to the effects of shallow water with small underkeel clearances and congestion in the anchorage that necessitated them to back and fill.

The third drill included a watch relief, the addition of current, and night arrival at a foreign port. The fourth drill incorporated a watch relief with a combination of equipment casualties, man overboard situations, wind and current effects, and traffic ships. Drill five was a day arrival, and drill six was a night departure, each with increasing levels of complexity. The same scenarios have been used since they were introduced in 1985. The mariner instructors found that the combined effects of limited nautical backgrounds, interpersonal dynamics associated with working as a team, officer-of-the-watch responsibilities, and the complexity of watchkeeping effectively prevented cadets from "programming" themselves in advance to do well in the course.

Different real and hybrid operating areas (that is, part real and part fictitious) were used for each exercise to guard against the masking of performance difficulties that might occur if trainees were to become familiar and at ease with the specific locale used in the scenario. In addition, the use of different operating areas exposed the trainees to differences in nautical charts and publications, the diversity of ports and approaches, and a range of operating conditions they could expect to encounter during actual service aboard merchant vessels.

The general type of graded scenarios and the operating areas simulated were:

- daytime arrival at the Panama Canal;
- daytime departure from Limon Bay, Panama;
- nighttime passage in Singapore Straits with a relief of the watch;
- daytime passage along the California coast with a relief of the watch;
- daytime arrival at Port International (fictitious port, San Clemente Island, California); and
- nighttime departure from Port International.

The use of real operating areas allowed the use of real nautical charts and publications. Hybrid operating areas (e.g., Port International) consisted of real operating areas with fictitious ports and fairways. These fictitious ports allowed use of real nautical charts and publications with a minimum of supplemental information for the artificial port and fairways. The instructors reported that the scenarios remained relatively constant in content over the 10-year period.

The Learning Environment

The cadets were formed into three- or four-person watch teams that remained together for the duration of the course. Team members alternated, playing the roles of officer of the watch, radar observer, helmsman, and navigator. The radar observer position was not filled in three-person watch teams. The cadet officer of the watch was placed in complete charge of the watch. The instructor did not venture onto the bridge except when special role playing was required (e.g., to simulate the presence of a master or marine pilot). Otherwise, the instructor monitored the drill and provided stimulation from the simulator control room, which included closed-circuit television video and voice monitoring and recording. Virtually the entire responsibility for performance and the results of the simulation were in the hands of the cadets. Drill number one was the first time a cadet experienced this level of operational responsibility.

According to the course's mariner instructors, the practice of placing cadets in complete charge of the watch aboard the simulated ship provided more opportunity for broad professional growth than training assignments aboard an actual ship. This effect was attributed to the absence of operational risk, the instructor's control of the operational scenarios that were experienced, and fair treatment during grading (discussed in the next section). Unlike training assignments aboard actual ships, there was no need for the master or officer of the watch to step in before a situation got out of hand (the circumstance that occurs when the safety of the vessel must necessarily be placed above learning objectives). The simulation allowed the cadet watchstanders to experience the full range of cause and effect relationships, and the results of those interactions, and to gain professional maturity through the constructive debriefing. The instructors could stop a simulation instantly if errant decisions so compromised watch team performance that individual confidence would have been eroded. Situations did not arise that required use of this capability.

Watch Team Performance Evaluation Criteria

Most ship-bridge simulator-based courses in which trainee performance is evaluated rely on instructor, or instructional team, subjective evaluation as to whether performance was satisfactory or unsatisfactory. In contrast, the USMMA watchkeeping course was graded according to very specific criteria. Figures F-1 and F-2 are examples of the cadet watch team grading and evaluation sheets. The overall grade for the course has three parts. One-third of the grade is based on watch team performance, one-third on each cadet's performance as officer of the watch, and one-third on a subjective assessment of cadet attitude during the graded drills.

CADET WATCH TEAM GRADING SHEET

DATE: 16 Feb 88 TEAM: A-1
WATCH OFFICER: LEE
NAVIGATOR: BROWN
RADAR OBS: BRIDGES
HELM: BURLEIGH

ARRIVAL CRISTOBAL Day [X] Night [ ]
(Anchorage in Limon Bay)
Range and bearing of anchor position arrived at from the planned anchor position:

TIME LET GO ⚓ 0643

EXECUTION
Total 40 Points - 2 Points Per Item

01. Compliance of Masters/Standing Orders 2
02. Proper preparation for Arrival 1
03. Proper internal communications 2
04. Proper vhf procedures 1
05. Master/Engine Room kept informed 1
06. ETA's maintained 2
07. Proper helm orders given 2
08. Frequency and method of pos'n fixing 2
09. Margins of Safety maintained 1
10. Optimum use of all navigation aids 2
11. Compliance with Port Regulations 2
12. Safe speed maintained at all times 2
13. Efficient visual lookout maintained 2
14. Anchoring properly prepared & executed 2
15. Optimum use of bridge personnel 2
16. Bell Book properly maintained 2
17. Log Book properly maintained 2
18. VHF log properly maintained 2
19. Anchored in correct anchorage 2
20. Ship satisfactorily maneuvered 2

EXECUTION SCORE 34

MONITORING
Total 20 Points - 2 points per Item

01. Track (Charted fixes and Pl) 1
02. Depths 2
03. Traffic 1
04. VHF 2
05. Helm 2
06. Instruments 1
07. Visibility/Weather 2
08. ETA's 2
09. Passing of information 2
10. Watch Officer 2

MONITORING SCORE 17

APPRAISAL & PLANNING
Total 30 Points - 2 Points Per Item

01. All relevant pubs. studied 2
02. Satisfactory plan on form 2
03. Tracks & courses on chart 2
04. Dangers and margins of safety marked 2
05. Tidal times and hts. calculated 2
06. Sufficient ukc/squat ascertained 2
07. Critical W/O marked correctly 2
08. ETA's and distances planned 2
09. VHF ch. noted and RP's marked 2
10. Frequency & method of fixing planned 1
11. Relevant Port Regulations considered 2
12. Weather expectations and forecasts 2
13. Ship's maneuvering capabilities considered 2
14. Contingency plans made 1
15. Effective anchoring plan made 2

APPRAISAL & PLANNING SCORE 27

ORGANIZATION & TEAMWORK
Total 10 Points - 5 points per Item

01. Watch officer composure 4
02. Teamwork 4

ORGANIZATION & TEAMWORK SCORE 8

SUMMARY
APPRAISAL & PLANNING (30) 27
EXECUTION (40) 34
MONITORING (20) 17
ORGANIZATION & TEAMWORK (10) 8

TOTAL POINTS (Out of 100)

AUTOMATIC DEDUCTIONS
1 Point for each minute late 0
15 Points for extremely poor navigation/grounding 0

ADJUSTED (FINAL) SCORE 86

COMMENTS: ONLY ONE STG PUMP CHK AFT STG NOT MANNED. ARRIVAL NOT SIGNALED. ZIG ZAG APPROACH THRU BKWTR. CONFUSION WITH OUTBOUND BRITANIC ENDEAVER & NO VHF TO AM. LANCER
PLANNING: EXCELLENT EXECUTION: GOOD
LET GO ⚓ WITH STERNWAY

INSTRUCTOR: [signature]

Beadon/Meurn/Sandberg 10/23/90

FIGURE F-1 Cadet watch team grading sheet. (Courtesy of Captain Robert Meurn, U.S. Merchant Marine Academy, and Captain Richard Beadon, Seamen's Church Institute).

APPENDIX F

EVALUATION OF __LEE__ SECTION # __115__
Circle
AS: (WATCH OFFICER) NAVIGATOR RADAR OBS. HELMSMAN

DATE __16 Feb 88__ TEAM __A-1__ SCENARIO __ARR CRISTOBAL__

EVALUATION CRITERIA	10 OS	9 VG:	8 G	7 S	5 P	0 UN-SAT	Comments
1. RELIABILITY (In class on time & CAORF by 0650)		✓					ARRIVED ON BRIDGE 5 MIN. PRIOR TO START
2. PREPAREDNESS (Fit & ready to assume duty)				✓			NAVIGATOR FORGOT BELL BOOK
3. PROFESSIONAL KNOWLEDGE (For duty assigned)	✓						
4. DECISIVENESS (Authoritative decision making)	✓						TOOK CHARGE OF WATCH
5. INITIATIVE (Self initiated activity)	✓						
6. FORCEFULNESS (Takes charge to carry out duty)	✓						
7. JUDGEMENT (Sound, mature thinking)		✓					DID NOT PLAN ABORT POSITION OR FOR CONTINGENCIES
8. COOPERATION (During debrief & with team members)			✓				A FEW EXCUSES DURING DEBRIEF BLAMED NAV.
9. RESPONSE DURING STRESS (Calm or excitable)		✓					
10. NAUTICAL TERMINOLOGY (Proper phraseology)				✓			POOR TERMINOLOGY IN LETTING GO ⚓
TOTAL	40	27	8	14			

GRADE __89__

Overall Comment (if any) __OVERALL EXCELLENT FOR FIRST GRADING EXERCISE__

Master Mariner

FIGURE F-2 Cadet watch team evaluation sheet. (Courtesy of Captain Robert Meurn, U.S. Merchant Marine Academy, and Captain Richard Beadon, Seamen's Church Institute).

Classroom sessions were not graded. Evaluation criteria for the graded drills had four elements:

- appraisal and planning,
- execution,
- monitoring and organization, and
- teamwork.

Grading began in the fifth week of the course with the first drill. Grading was tailored for each scenario. During the 10-year period of the case study for this appendix, automatic deductions were consistently applied for watch team failure to comply with specific training objectives. The grades for watch team performance were averaged over the five or six drills for a composite grade. Each individual's drill was assessed subjectively by the lead instructor for that watch team. A composite grade for attitude was derived by averaging the scores for the complete series of drills.

Some of the instructors who conducted the course have left, although some have been involved in the course throughout the 10-year case study period. All instructors had prior experience as master and had received training in the use of simulators for instruction. Considerable care was taken by the instructional staff to ensure consistency and evenness in their grading of cadet watch team performance. Measures to ensure consistency and evenness across cadet watch teams and courses involved strict adherence to the course's overall aims and scenario-specific objectives.

Analysis of Course Performance Data

The data were tabulated and compared for each course. The average increase in absolute score was about 10 percent over 5 drills and 11 percent for 6 drills. Several courses resulted in slight decreases in absolute scores over the period. Two cadet teams received no points for one drill. These two scores were not included in the tabulated data because the drills were not conducted and the scores were assigned for reasons other than performance during simulation.

All data were then placed on a line chart to graphically convey the density of the grade distribution (Figure F-3). A modest upward trend in performance is suggested by the dark band at the top of the chart. The density graph, by visual interpretation, revealed considerable variability in assigned scores for the first two drills. The observed variability in scores decreased markedly beginning with the third drill and continued a gradual but steady decline thereafter, as indicated by the calculated standard deviation (Figure F-4). The increase in assigned scores and decrease in variability among teams' performances coincided with increases in the degree of difficulty, observed improvements in watch team cohesion, increased familiarity with the "mechanics" of effective watchstanding, and acclimation to the high level of personal responsibility associated with duties as

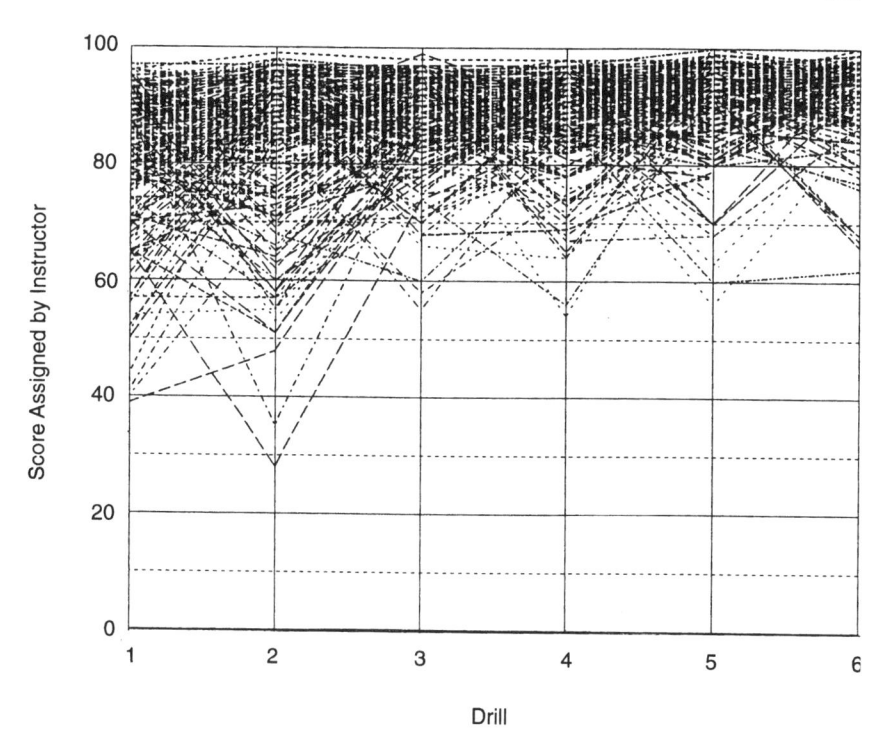

FIGURE F-3 Scores achieved by 233 cadet watchkeeping teams undergoing a simulation-based watchkeeping course during the period 1985–1994.

officer of the watch. The instructors attributed the increased spread of scores in the second drill to the cadet watch team's lack of appreciation for the increase in the diameter of the turning circle associated with shallow water and small underkeel clearances and the implications of these effects to maneuvering, in this case, the need to back and fill the M/V *Capella* in order to safely depart a crowded anchorage.

Data were not available that might enable assessment of the degree to which the high variability in the scores for the first two simulation drills might have been influenced by differences in watchkeeping practices aboard the Academy's training vessel and merchant ships used during cadet training periods that preceded the simulator-based training. A plot of the standard deviation for each drill, without regard for the increasing degree of difficulty (Figure F-4) suggests that the overall improvement was gradual but steady after the second drill. The gap between the upper and lower grades suggest that several additional drills could potentially be used to optimize the training program for some cadet watch teams.

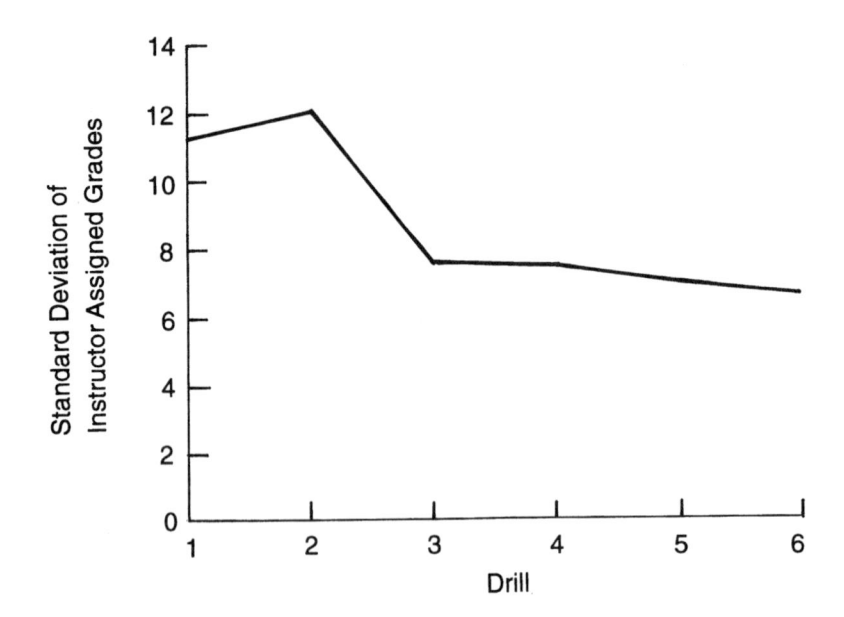

FIGURE F-4 Plot of standard deviation for simulation-based cadet watchstanding course.

Figure F-5 is a plot of the average absolute scores per drill for all cadet watch teams and a weighted average of scores. The weighted average was obtained by multiplying each average absolute score by 110 percent to reflect the estimated 10 percent increase in difficulty with successive drills. The weighted average increase in score was about 54 percent over 5 drills and 67 percent for 6 drills.

Unlike actual operations, the instructors were able to control the operating environment to maximize the concurrent development of basic watchstanding procedures (i.e., the mechanics of standing an effective watch), bridge team coordination, navigation practices, and problem solving. The increase in both the average and weighted average scores does not mean that a similar result could not be obtained through the traditional development of watchstanding capabilities. The lack of control over the operational scenarios that are experienced would, in the committee's experience, require more time to obtain a similar result. The lack of scenario control would also be unlikely to provide either the full range of operational scenarios used in simulation or the exposure to the range of ports and operating conditions that were experienced during the CAORF simulations.

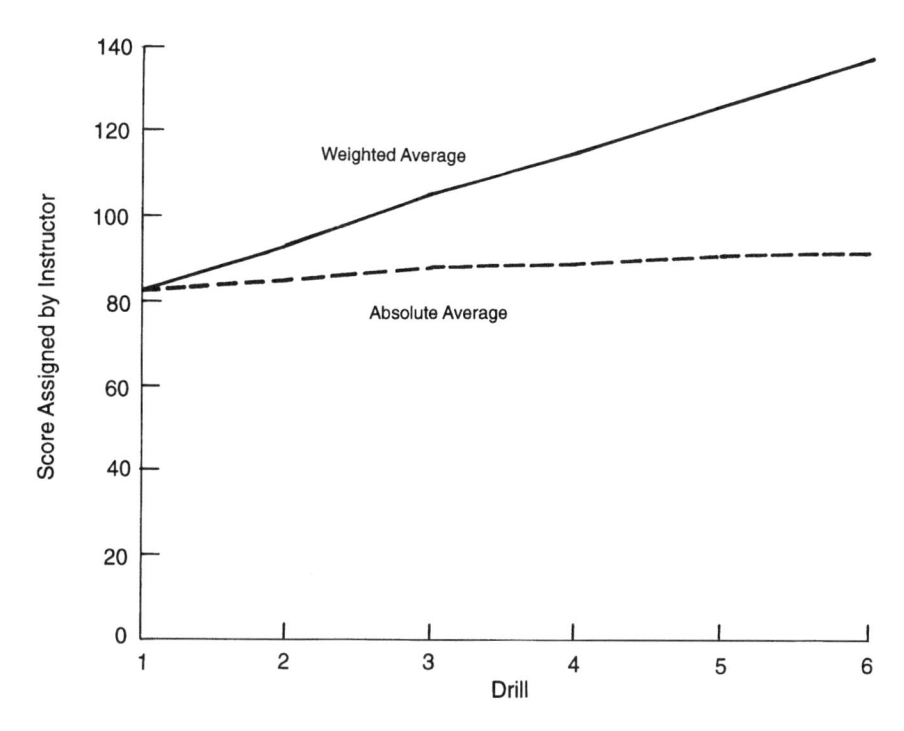

FIGURE F-5 Average and weighted average scores per drill of simulation-based watch-keeping training at the U.S. Merchant Marine Academy.

Reinforcement of Simulator-Based Training

No data were available to determine if overall performance could have been enhanced by a more systematic correlation of the simulator-based course curricula and school ship or merchant ship training criteria. A consistent and correlated approach would, however, progressively reinforce the best procedures and practices.

There was no formal program to monitor performance of trainees following the graduation or self-study programs to help reinforce lessons learned from the simulator-based training. Cadets were invited to correspond with the instructors following their first watches at sea as officer of the watch; however there is little anecdotal information of this type to enable an assessment of transfer effectiveness. The responses that were received are encouraging (see Box F-1). It appears from the responses that transfer has occurred in varying degrees, although the mariner instructors believe that optimal transfer has been inhibited by variability in bridge resource management practices aboard ship.

> **BOX F-1**
> **Third Mate Observations on Value of**
> **Ship-Bridge Simulation Cadet Watchkeeping Course,**
> **U.S. Merchant Marine Academy**
>
> I'm sailing as relief third mate . . . and your teaching on Rules of the Road has made it easier to cope with the butterflies involved with a third mate's first watches. I find myself evaluating the situation systematically.
>
> Jorge J. Viso, September 30, 1985
>
> I did not assume my first actual bridge watch until we were well out to sea on our way to Valdez, Alaska. By this time, I had made good use of the opportunity to familiarize myself with the bridge, radars, steering gear, etc. Still, I was a little apprehensive when I stepped on the bridge at 1930 sharp. After the Chief Mate gave me a few encouraging words, he left the bridge, and I had officially assumed my first bridge watch. The captain was not on the bridge, but he had told me that if I had any problems to call him. The first watch was uneventful. There was some traffic, but nothing came closer than six miles. I have to give CAORF a lot of credit for preparing me for bridge watchstanding. After some initial jitters, I felt very comfortable and confident. In fact, I was much more nervous standing my first watch on the *Capella*. After going through CAORF, I feel that I am ready for anything, steering failures, loss of plant, etc. Since that first watch, I have had three watches which required maneuvering for traffic situations. For all three situations, I spoke with the other ships over the VHF regarding the course of action we were going to follow. One was a starboard to starboard passing. On another occasion, we lost one of the boilers. The ship slowed from 80.2 revolutions per minute to 60. I was in the chart room filling out a weather report when I noticed the ever present rattling stop. The first place I looked was the RPM indicator which was steadily dropping. I had the helmsman go to hand steering, looked for traffic on both radars and visual, and was about to call the Captain when he came up to the bridge with a weather map. He had already known what was going on and said the engineers had been having problems with the boiler. The same type of situation had happened to me on the *Capella*.
>
> Robert Lenahan, July 29, 1986
>
> It is now only one and one-half months after graduation and I honestly feel comfortable "in control." I had the wonderful experience of passing through the Florida Straits in a "special circumstance" very similar to a CAORF simulation. . . . I was amazed how smoothly I handled it. CAORF certainly does boost one's confidence.
>
> Robert E. Munchbach, August 4, 1993

APPENDIX F
259

CASE STUDY TWO
SIMULATOR-AIDED MARINE PILOT DEVELOPMENT,
PANAMA CANAL[1]

Marine simulation has become an important training aid in the development of Panama Canal pilots. This case study describes pilotage and pilot development and use of simulators.

Panama Canal Pilotage and Pilot Development

The Panama Canal is still operated by the U.S. government. Panama Canal pilots are currently federal employees, although this status is in transition under the Panama Canal Treaty. The Panama Canal Company, the operating arm of the Panama Canal Commission (PCC), employs 242 Panama Canal pilots and pilot trainees. The PCC, as a federal commission, established pilot development requirements that generally follow the federal pilotage program with respect to organization and licensing. A license as master (unlimited oceans) had been an entry-level requirement, but was reduced to aid in meeting requirements of the treaty, as discussed in this case study. There are eight steps (grades) of qualified pilots. After a candidate becomes a qualified step one pilot, advancement from smaller to larger ships is based on time served and other criteria. It takes about eight years to become a fully qualified step eight pilot. The only restriction at step eight is that a pilot needs an observation ride on a submarine before serving as a pilot aboard a submarine.

A structured pilot apprentice program is used for individuals qualified to enter the pilot development program. Qualified pilot examiners conduct check rides as part of initial licensing. Check rides are also used for upgrading purposes during the first two years of progressive advancement. The check rides are done by qualified Panama Canal pilots who serve as evaluators on a voluntary basis. In the past, check rides were qualitative, although some quantitative measures have begun to be applied. The program is technically guided by a substantial training manual, *The Panama Canal Pilot* (Markham, 1990). The program and specific pilot development criteria are described in *The Panama Canal Pilot Training Programs* (PCC, 1993).

Changes in Pilot Development Motivated
by the Panama Canal Treaty

The pilot development program changed dramatically in response to the Panama Canal Treaty. The treaty requires that all pilots be of Panamanian nationality once the Panama Canal is returned to Panama. This requirement created

---

[1]Except where indicated in the text, this case study is based on a presentation to the committee by Captain S. Orlando Allard, Chief Training Officer, Panama Canal Commission, November 8, 1993.

a significant training challenge because when the treaty was signed, no Panamanians were qualified to enter the pilot training program. This approach to the problem included extensive use of ship-bridge simulators as part of a revised training program.

Anticipating the treaty, the PCC had begun a long-range pilot development program before the treaty was signed. Although the marine license requirement was lowered to enable entry by tugboat masters and ships' officers, this move did not resolve the deficit in pilot candidates of Panamanian nationality. The Panama Canal Company tugboats were not operated by Panamanian citizens.

To encourage Panamanians to enter the pilot development program and to expand the pool of individuals to choose from, the PCC established a preparatory program. Panamanian citizens were placed aboard PCC tugboats for on-the-job professional development. These individuals were progressively advanced to tugboat master over five years. After completing this preparatory training, qualified Panamanians were moved into the pilot training program. Even with this program in full operation, there were insufficient Panamanians available that were qualified to enter the pilot development program, so a pilot "understudy" program was developed and implemented. The understudy program consists of nautical schooling and two years of licensed service as third mate aboard vessels of 1,600 gross tons.

In setting up the apprenticeship pilot-in-training (PIT) program, it was necessary to look at the specific tasks required of a pilot. Understanding what a pilot needs to know and be capable of doing is especially important in the Panama Canal because the pilot assumes a greater degree of control over vessel maneuvering than occurs under pilotage common to most seaports (NRC, 1994). English is taught as part of the PIT program to improve the English language skills of pilot apprentices. Training in the pilot understudy program is highly individualized. Between 20 and 25 individuals are in the program at any one time. No more than six individuals are accepted from any of the three avenues to enter the PIT program. This limitation was imposed to allow the training program to be tailored to each individual. Each candidate, however, is required to go through the same progression of check rides and series of examinations in qualifying. Candidates are limited to a maximum of six solo transits aboard small vessels during the PIT program. The rest of the transits must be with a qualified pilot.

Training consists of three segments, with a written exam at the end of each segment. The candidate must draw charts of the pilot routes. Each trainee must also present a graduation project, a type of thesis. Projects are assigned to help pilots. The PCC also has a "transit advisor" position for vessels under 65 feet. Trainees are placed aboard these small vessels to acquire additional practical experience, but their role is strictly that of an advisor. As a practical matter, the trainee handles the vessel, and the vessel's master will not take control back unless the trainee advisor experiences a problem.

Application of Simulators in
Panama Canal Pilot Training

The PCC has had considerable experience with marine simulators. Computer-based shiphandling simulation was used extensively in the design of improvements for the Galliard Cut. A number of Panama Canal pilots participated in this applied research (NRC, 1992). Under labor-management contractual arrangements, the PCC is not able to require the use of simulators for the training of licensed Panama Canal pilots or for performance assessments of licensed pilots. Voluntary training in shiphandling has, however, been used for many years. Before acquiring its own simulator capabilities, the PCC choose facilities that had former Panama Canal pilots on staff as instructors. The first simulator PCC installed was a small ship-bridge simulator used for training of licensed pilots. The PCC now also has a full-mission ship-bridge simulator, which has become an essential training resource to assist in qualifying Panamanian citizens as pilots. In using simulators, the objective of the PCC is to enhance the training provided, not to replace hands-on experience.

The PCC has established a series of workshops that use simulators. Hands-on training is emphasized. Trainees moderate the workshops. The small simulator is used in the workshops; the large simulator is used only for actual training. In general, the senior pilots have not been exposed to this training. Because of the compensation structure, pilots are generally unwilling to come in voluntarily to participate unless they receive credit for transits—the basis on which compensation is calculated. As a result, simulator-based training has primarily involved junior personnel.

Motivations for Increased Use of Simulators

In addition to treaty requirements, a number of other factors have placed pressure on the professional development of pilot apprentices. More pilots are being drawn from the pool of inexperienced mariners who lack the extensive sea-service experience associated with mariners who came into the program in previous years. In addition, the average age of pilots is going down. Age is an important consideration, because Panama Canal pilots need a strong command presence.

Because of the great degree of control over vessel operation exercised by Panama Canal pilots under PCC rules, considerable professional maturity is needed. To aid in the development of command presence, professional maturity, and piloting knowledge and skills, the PCC established a mentor program. A trainee accompanies a senior pilot during an entire month. The PCC has also established a program in which two trainees accompany a lock approach pilot and participate in a large number of approaches and lockings. During most of these trips, a trainee is an observer. Pilots do not go into the docks as much as in the past at the terminal ends (the docks are now operated by the Panamanian government rather than by the commission).

The trend toward larger ships has adversely affected the number of smaller ships available as training platforms. This professional development problem is significant and long term because the refinement of piloting skills has relied heavily on progressive advancement. The use of simulators is one way the commission has attempted to adjust to these changing trends and operating conditions. Each pilot candidate is required to participate in five days of training on the large simulator. Sea-time equivalency is granted on a 1 to 1 ratio. Simulation has become more important in developing command presence because, as the composition of the pilot cadre changes in response to treaty requirements, there are fewer senior pilots with whom to ride.

Simulator-Based Training Applications

A pilot has to be capable of spontaneously anticipating what will happen with respect to operational scenarios and ship behavior. Pilots must also be capable of mentally preplanning the transit. For example, the trainee should expect certain hydrodynamic effects under certain conditions. If these effects are not experienced or are different than anticipated, then the trainee should inquire why the effect did not appear or was different and adjust accordingly. Although ship-bridge simulators have limitations, it has been a most useful resource for developing anticipatory abilities.

Simulation is used to provide exposure to tasks and situations such as dockings, moorings, anchoring, and restricted visibility in Galliard Cut, which have not always been available in the canal. Simulation is also used to practice piloting in meeting situations and helps fill the gap created by the lesser number of suitably sized ships as training platforms. Simulation provides a capability for repetition of tasks and emergency response exercises. Scenarios from past canal accidents are used in the emergency response exercises. The training staff has not attempted to replicate the real world for most aspects of pilot development. The observation opportunities in the Panama Canal are such that there is no need to fully replicate real-world activity.

The present training resources are limited compared to the number of pilot candidates undergoing training. There are a limited number of instructors, necessitating the use of the same instructor for much of the training. In the experience of the instructors, this situation reduces the training value somewhat. The training staff, however, is able to cover procedures that need to be followed to respond effectively. To establish a realistic master-pilot relationship, it is effective in the training program for a Panama Canal pilot to play the role of master during simulation. It also is important to have a pilot or someone familiar with the canal at the control station to observe and assess the simulation. The role-playing opportunity also helps build support among the licensed pilots for simulator-based training.

The training includes all of the tasks in simulation that would occur during a real transit. Close quarters maneuvering has not been emphasized in the training because of the difficulty in producing sufficiently accurate hydrodynamic interactions. The decision making needed in meetings and other ship-to-ship interactions are emphasized. The canal also offers substantial opportunities to observe close maneuvers through lock approaches. These opportunities, in effect, serve as real-time simulations for trainees. High-fidelity hydrodynamics are not considered important for many simulation exercises, such as those in which communications are the primary training objective. The training staff uses many short exercises with very specific objectives. Some communications exercises are conducted to address language difficulties.

The training staff has found videotapes of the simulation exercises to be very useful training, debriefing, and assessment aids. The exercise can be replayed in plain view or through the windows. The debriefing process is emphasized as part of the simulator-based training. Simulation has not been used as a testing tool, although there are plans to use simulation for testing for some aspects of training.

Presently, there are no measures of effectiveness for simulator-based training. Such measures have been very difficult to develop. Observations of trainee performance and capabilities are made over an extended period to develop a more complete profile of an individual's preparedness. Box F-2 summarizes a Panama pilot's observations on the value of ship-bridge simulation training. The professional credibility of the evaluator is also important to this process in the PCC program.

CASE STUDY THREE
PORT AND VESSEL FAMILIARIZATION, AMERICAN PRESIDENT LINES[2]

American President Lines (APL) is a U.S. shipping company. Its container ships presently fly the U.S. flag (although there are plans to reflag some ships). APL has 42 permanent masters (i.e., captains permanently assigned by their unions to specific APL ships). APL has been a major sponsor of ship-bridge simulator-based training and the use of simulators for assessing vessel operating criteria. The company also conducts onboard audits of individual performance. The company began sponsoring bridge resource management training to provide professional training between onboard audits. The company is considering bringing the entire bridge team in for training, rather than just the senior officers.

Perception of Simulator-Based Training

Simulation provides an opportunity to correct bridge team management deficiencies and problems in personal styles without interfering with and jeopardizing

---

[2]The material presented in this case study is based on a presentation to the committee by Captain Saunders Jones, American President Lines, November 8, 1993.

> **BOX F-2**
> **Observations of Panama Canal Pilots on the**
> **Value of Ship-Bridge Simulation Training**
>
> "Getting the chance to practice maneuvers in Canal anchorage areas in the simulator has allowed me to develop more confidence while maneuvering in crowded anchorages, especially in Limon Bay anchorage." (Pilot understudy)
>
> "My communications skills have improved since I have trained on the simulator. Communications with other vessels, marine traffic, and signal stations has become clearer." (Pilot understudy)
>
> "Backing and filling maneuvers in closed quarters have become a lot smoother since I have been able to practice them in the simulator." (Pilot understudy)
>
> "Maneuvering container vessels with strong winds abeam is one of the maneuvers that I have been able to practice in the simulator that has given me an edge when I have had to do it for real, especially in the approaches to Gutan Locks." (Licensed pilot, two years experience)
>
> "I have been able to experience and better understand the interaction between vessels when passing each other." (Limited pilot)
>
> "I had the opportunity to practice docking situations at dock 16B Cristobal before I actually had to do it for the first time. The simulator helped me obtain a better understanding of the situation." (Limited pilot)
>
> "The vessel I was piloting suffered an engine failure. Since I had practiced exercises in the simulator where this type of situation occurred, my response to the real event was quicker and more confident." (Panama Canal pilot)
>
> "While transiting the narrowest part of the Canal, the Culebre Cut, heavy fog set down and the visibility was reduced to almost nil. Having practiced this very scenario in the simulator, I was able to handle this situation in a calm and relaxed manner." (Panama Canal pilot)
>
> "The main disadvantage I have experienced in the simulator is the perspective looking from the bridge. It has taken me some time to get used to it, and it differs somewhat from what you see on board vessels." (Pilot understudy)
>
> "Simulator training has helped me understand and set my priorities when making specific maneuvers." (Pilot in training)

real-world operations. For example, some masters overload themselves and do not effectively distribute tasks. This situation cannot be corrected on the real bridge because it is too dangerous. Simulation is not a replacement for shipboard experience, and especially not a replacement of unusual events.

Simulator-based training, in APL's experience, broadens the student's window of experience and accelerates his or her move up the learning curve. Multiple exposure, repetitions, and immediate feedback are key benefits. Simulation provides the capability to experiment with alternatives, to see what works and what does not, and to do this without risk. Simulation is also considered to be a valuable medium for use in remedial training, skill development and enhancement, passage planning, port familiarization, shiphandling, familiarization with new vessels, and division of tasks in the ship-bridge operating environment.

Vessel and Port Familiarization

APL has used simulation to selectively introduce its personnel to new equipment and new ship types before the ships enter service. For example, the company sponsored simulated port entries of its new C10 container ships before they entered service in Oakland, California, and in Dutch Harbor, Alaska. In each case, a cadre of local pilots participated in the simulations at APL expense to assist the local pilots in preparing to handle these transits.

In the case of Dutch Harbor, the company had very little operational experience with extremely large container vessels at this new facility in a new location. In particular, the Dutch Harbor area is prone to catalatic wind conditions (known in the region as "williwaws") which occur during certain atmospheric conditions. The company used ship-bridge simulation to develop the operating criteria and standards to be used to guide actual operations. Because they had been prepared in advance for the conditions that were actually encountered, APL masters and local pilots were able to reduce the number of missed port calls and do more in actual operations. The simulation opportunities were particularly beneficial for the pilots.

A shift in state pilotage district boundaries resulted in pilotage services being provided by a new group who had less experience with large ships. Fifteen pilots trained during three training courses. The program was evaluated by the company and the pilots as extremely successful. A very fast learning curve was demonstrated, and the simulator-based training helped establish a working relationship among the pilots and APL's ships' officers.

CASE STUDY FOUR
COORDINATED TUG SIMULATOR-BASED ONBOARD DRILLS, MORANIA OIL TANKER CORPORATION[3]

The Training Program

Representatives of the Morania Oil Tanker Corporation provided an overview of Morania's training program. The company operates about two dozen towing vessels, with both licensed and unlicensed personnel. The unlicensed

---

[3]The material presented in this case study is based on presentations to the committee by James Sweeney, Morania Oil Tanker Corporation, and Captain Herb Groh, marine training consultant, November 8, 1993.

personnel are important and are now part of the training process, and the classroom has been placed on board the vessel.

The company examined its operations and determined that it had significant training needs. The company was aware that a simulator-based training program for towing-vessel operations had been initiated by another towing industry company, Maritrans, a decade earlier. The results of the Maritrans program were encouraging and contributed to Morania's decision to sponsor simulator-based training for its licensed operators.

The company's initial objective was to find out what their operators knew and what they did not know, especially with respect to gaps in knowledge and situational awareness. The company decided that it would not provide simulator-based instruction in boathandling, because boathandling was already a fundamental component of learning by experience. Initially, the company sponsored training for its tug operators and for the operators of tugs it had chartered. A company representative attended the ship-bridge simulator-based training to gain a sense of operator receptivity to the training.

About a year after the initial simulator-based training, the company decided to continue with the training program. It was believed, however, that the vessel personnel were not ready for additional simulator-based training. The company, therefore, opted for an onboard audit of what had been learned and applied from the simulator. The onboard audit was conducted in the year following training. Initially, the company's port captains and senior engineer conducted the audits. These individuals, however, were too close to the operators, and the results were not sufficiently objective. The company decided there was a need for a credible third party to conduct the audits and contracted with a marine training consultant, who had also served as the simulation instructor for the computer-based training sessions. All vessels have now been audited. In the second year following training, the company decided to have their training consultant conduct onboard drills or "real-time" simulation.

On Board Training and Performance Evaluations

The integrated tug-barge units are about 600 feet long, with a speed of approximately 6 knots fully loaded and 10 knots light. The tug watch officer does not have excess power to maneuver the vessel. They use their experience to get out of tough situations.

As part of the instruction, the instructor discusses what needs to be done during different scenarios. After the drills, the instructor conducts an onboard debriefing. To instruct all crew members, since they are on port and starboard schedules, the riding instructor visits each unit twice. The company sends him anywhere the unit is located to complete the training. After the onboard training, the instructor files extensive reports on training results.

Onboard training is considered to be an outstanding way to find out what is going on aboard a vessel. Because the same instructor was used during simulation and onboard instruction, an instructor-student relationship had already been established. In addition, the instructor was aware of what to expect from each individual, based on his or her participation in the simulator-based training course. The instructor was not on board to instruct in or evaluate shiphandling. When on board, the instructor evaluated the performance of the vessel (and its crew) using his nautical expertise as a reference (the instructor's credentials included 50 years of maritime service as master and pilot). The instructor reported that in his experience, he found nothing better than looking over the operator's shoulder. (Of course, the operators were not always receptive to this situation.)

The testing and validation of simulations can be expensive. Once the data and adjustments are in the program, these aspects are basically not transportable. The price of simulation probably will not come down until economies of scale are achieved through more and wider use. Although insurance underwriters base rates on costs to the insurance industry, the company has been able to negotiate some rate reduction as a result of its documentation of training results and improved operator safety performance. Over the last three years, Morania has had a reduction in personnel and indemnity and hull insurance rates, which it attributes in large measure to improved safety performance as a result of the training program.

CASE STUDY FIVE
U.S. COAST GUARD-APPROVED SIMULATOR-BASED MASTER'S LEVEL PROFICIENCY COURSE[4]

U.S. Coast Guard Course Approval

In late 1994, for the first time, the U.S. Coast Guard (USCG) approved the use of a ship-bridge simulator-based course as an alternative to the standard multiple-choice examination. After several years of planning and development, SIMSHIP Corporation made an unsolicited proposal to the USCG requesting approval of simulator-based testing as a substitute for portions of the master (unlimited oceans) written examination for which ship-bridge simulation was suitable. A full-mission ship-bridge simulator was proposed as the testing platform.

The USCG is interested in encouraging professional training to improve the practical preparation of mariners for service. The agency did not accept a substitute

---

[4]The material presented in this case study is a composite of unpublished information provided by Captain Frank Seitz, SIMSHIP Corporation, Frank Flyntz, U.S. Coast Guard, and Robert Spears, U.S. Coast Guard.

of a more costly testing platform for its license examination. It did, however, indicate that it was interested in an alternative that would combine training and performance testing. The agency ultimately accepted and approved a combined training and testing course using a full-mission ship-bridge simulator as a voluntary alternative to the entire master (unlimited oceans) license examination. The USCG believes that the course will go above and beyond what is currently required in the license examination and that the approach will increase individual abilities to effectively apply knowledge.

Approval of the training-testing course concept relied on the USCG's interpretation of existing, enabling authority for course approvals found in 46 CFR 10. According to the agency, to remain within the scope of enabling authority, testing must be an integral part of the course. In addition, the approval process must be applied to the curriculum, to the facility at which the course will be conducted, and to the instructors and evaluators.

The initial course approval was for a two-year period. If the approach is successful, the course would be regulated thereafter under the USCG's standard five-year renewal policy. Since the course is approved as a substitute for the license examination, the USCG is not granting any credit for sea time. Participation in the course, which is being held at simulator facilities operated by the American Maritime Officers (AMO), is being subsidized by AMO for union members. License candidates not belonging to the union, but wishing to participate, would be required to pay all costs or the costs could be borne by their employers.

Official Oversight

A significant difference from current practice, in which USCG license examiners conduct the testing, is that representatives of SIMSHIP Corporation conduct both the training and testing, with USCG oversight. A USCG representative is scheduled to attend at least two of the first three testing courses. The USCG has a goal of overseeing the course two out of every three offerings for the first two years. Thereafter, the frequency of USCG visits to spot check the training and testing processes will depend on the quality of the course. If high quality is obtained, the agency anticipates that its visits and checks would be on a part-time basis. Nevertheless, the USCG plans to conduct checks periodically.

The USCG, in approving the course, stipulated that SIMSHIP Corporation train a USCG representative (with either commercial or naval marine experience) for the oversight role. The objective of the training is to prepare the agency representative sufficiently so that the agency is capable of determining compliance with terms and conditions of the course-approval criteria. Additional stipulations are as follows:

- The course will be used as a substitute for the master (unlimited oceans) examination only.
- The lead instructor and evaluator must hold a master's (unlimited oceans)

license and have served as master for one year on that license on a ship of 5,000 gross tons or greater.
- Individuals with lesser qualifications could assist the lead instructor or evaluator, but a marine license is mandatory.
- Each instructor and evaluator is required to have training in the use of simulation and in basic teaching and evaluation techniques.
- No individual can serve as both an instructor and evaluator during any one course.
- Other students or license candidates may not role-play during performance evaluation components of the course.
- The training and testing course will be open at any and all times to USCG visitation, participation, and inspection.

There are no industrywide or USCG standards for the training of instructors or evaluators. The training may be provided "in-house" or obtained from other sources. An in-house "train-the-trainer" course has been developed by an expert in training systems and technology and is in use for training instructors in other courses offered by SIMSHIP Corporation.

The USCG did not require formal validation of the simulator or the simulation as part of the course-approval process. The agency relied on SIMSHIP Corporation to self-validate the course. This self-validation was done through a structured series of live trials using volunteer test subjects, including chief mates, master mariners, and marine pilots. SIMSHIP's interdisciplinary staff, including a hydrodynamicist, master mariners, and individuals with extensive experience in the use of simulator-based mariner training, were also involved. A description of the validation process was not required prior to course approval. As part of the approval process, USCG representatives holding relevant marine licenses witnessed the dedicated trials and examination portions of the course.

The USCG has not established a monitoring program to determine the effectiveness of the training and evaluation course with respect to subsequent actual performance. As a quality control measure, the USCG will require that each candidate take one of the multiple-choice testing modules. The module will be selected at random and administered at a USCG regional examination center by a USCG license examiner. The candidate will be issued his or her license regardless of the outcome of the written examination. The USCG plans to retain and analyze the results of the random testing and will consider the results in determining whether to renew the course approval.

Course Description and Process

The approved training and testing consists of a two-week course. The approval documents stated that a maximum of four license candidates will participate in any given course. Each element of the current license examination is featured in the course curriculum. Heavy emphasis is placed on bridge resource management and maneuvering ships in restricted waters. Every course topic is required to be at the master level. Student performance on examinations will be tracked through check-off lists during training and testing.

Training and testing will be conducted separately. The first week is devoted to training, which will address the subjects to be examined on the simulator. Students must also be prepared to take written examinations on topics that are not going to be examined, using the simulator as the evaluation platform. The second week is devoted to testing for all elements of the license examination. The instructors involved in training during the first week are precluded from serving as evaluators the second week. The USCG's stipulations do not preclude instructors from being involved in role playing.

The testing will include practical demonstrations for communications, chart work, bridge resource management, and situational awareness. Ship-bridge simulation will be used only for those elements of the testing for which it is suited (e.g., bridge resource management, rules of the road, shiphandling). A generic port is used for the simulations. Written tests will continue to be used for most other elements of the examination, including ship's business, and will be drawn from the existing pool of USCG-maintained, multiple-choice license examination questions.

The scoring of performance demonstrations will be done on a pass-fail basis and measured against specific, weighted criteria. This approach was adopted to remove subjective judgment as much as practicable from the assessment component of the course. The determination will rely on the professional expertise of the evaluators. The USCG has not established specific criteria for use as assessment benchmarks. The agency believes that course materials and any criteria developed and used by SIMSHIP Corporation are proprietary, even though the course is approved as a replacement for an official licensing examination.

License Administration

On successful completion of the course, license candidates will be issued a certificate that can be presented to the responsible USCG licensing official. An individual who does not pass the training and testing course has the option of (1) retaking the course or (2) taking the USCG's standard multiple-choice examination. If retaking the course, the candidate would have to return at a later date. The length of time would vary, depending on the degree to which the individual's participation fell short of performance expectations. Under current criteria, the maximum number of times an individual may take the course is two. During the course, a participant can fail and retake one of the modules. Any trainee failing two of the modules is "washed out" of the course. Individuals who do not successfully complete the course can choose to take the written USCG examination without any time delay. In this situation, the USCG would view the course as nonrequired training.

REFERENCES

Markham, G.A. 1990. The Panama Canal Pilot. Balboa, Panama: Panama Canal Commission.

Meurn, R.J. 1990. Watchstanding Guide for the Merchant Officer. Centreville, Maryland: Cornell Maritime Press.

NRC (National Research Council). 1992. Shiphandling Simulation: Application to Waterway Design. W. Webster, ed. Committee on Shiphandling Simulation, Marine Board. Washington, D.C.: National Academy Press.

NRC (National Research Council). 1994. Minding the Helm: Marine Navigation and Piloting. Committee on Advances in Navigation and Piloting, Marine Board. Washington, D.C.: National Academy Press.

PCC (Panama Canal Commission). 1993. The Panama Canal Pilot Training Programs. Unbound compilation of pilot training program materials. Balboa, Republic of Panama: PCC.

APPENDIX

G

Microcomputer Desktop Simulation

This appendix discusses the background and uses of microcomputer desktop simulators. In this appendix, a basic microcomputer desktop simulator consists of a microcomputer with hard drive, a single cathode ray tube color monitor, keyboard, an auxiliary control device such as a mouse or trackball, data input-output capabilities, and simulation software. The data input-output capabilities would consist of one drive or a combination of floppy or other types of drives, possibly a network connection (e.g., for microcomputers linked into a classroom training system), and "read-only" devices such as a CD-ROM. This envisioned basic configuration would be analogous to personal computer systems found in many homes, businesses, and aboard many ships. The microcomputer workstation would be configured for individual use, although individuals might be linked to an instructor console. More-elaborate desktop systems might include such features as specially designed data-entry devices (or consoles) or multiple monitors.

INTEREST IN DESKTOP SIMULATIONS

The rapid increase in the computational power and videographics capabilities of microcomputers has stimulated interest within the marine community in using these capabilities to bring simulators onto ships' bridges and into the classroom and the home. The U.S. Coast Guard (USCG) is interested in understanding whether and how microcomputer simulators could be used as a supplement or an alternative to ship-bridge simulators. The agency views microcomputer-based simulators as a possible reduced-cost option for applications to mariner

APPENDIX G 273

training and performance evaluations. The agency currently requires radar observer certification which uses radar simulators as the training medium. Movement toward using simulators, including desktop simulators, in marine licensing was recommended by an internal USCG focus group to provide a competency-based, rather than knowledge-based, license process (Anderson et al., 1993).

There is a growing belief that it might be possible to more accurately and completely assess a license applicant's ability to apply knowledge and skills for some tasks using desktop simulators rather than multiple-choice written examinations. If desktop simulators prove to be feasible, practical, and suitable with respect to training or licensing objectives for broad application in the professional development process, the reduced cost could potentially lead to wider availability of simulator-based training, performance evaluations, and licensing. Such a development would constitute a substantial change relative to the current practice of multiple-choice written examinations (ECO, 1987).

ASSESSMENT FACTORS

The following topics should be considered in determining the suitability of desktop simulations:

- the technical and instructional state of practice in microcomputer desktop simulators;
- the research and development basis for using microcomputer simulations in marine training and licensing;
- the possible use and applications of instructional design process;
- the training potential of microcomputer desktop simulators, including the ability to produce user behavior that would occur during actual operations and the potential for developing and retaining knowledge and skills;
- the changes to marine licensing recommended by the internal USCG study group relative to microcomputer simulations (Anderson et al., 1993);
- the potential of microcomputer simulators for reinforcing skills between scheduled, structured training courses;
- the quality of learning in controlled and self-instruction training environments;
- simulator evaluation methods and their applicability;
- diagnostic capabilities of simulators;
- characterization of trainee populations, tasks, and functions for which microcomputer simulator applications may be suitable;
- suitability for use for direct or indirect support of actual operations;
- need for simulator and simulation validation;
- ability of the simulator to be user-friendly; and
- cost effectiveness.

GENERAL STATE OF PRACTICE OF MICROCOMPUTER DESKTOP SIMULATORS

Computer-assisted learning has been used for some time by a number of organizations that offer license-preparation courses. Available courseware includes tutorials to aid in the acquisition of knowledge and to practice responding to questions in the multiple-choice format. There is a growing library of simulator software designed for marine applications. Desktop training simulations are commercially available for general navigation, radar navigation, piloting, ship-handling, maneuvering, automatic radar plotting aids, rules-of-the-road training, port entry, and the global maritime distress safety system. Some of these software packages are already being used to some extent for training and simulation of port entries in the classroom and aboard some ships; other packages are undergoing field evaluations. Because of computational requirements and presentation fidelity needs, the available software requires or works best with higher-level microprocessors and videographics array or super videographics array color monitors.

The technological capability also exists to emulate electronic navigation equipment, such as radars and automatic radar plotting aids, at modest cost using microcomputer hardware and software. Because such emulations are driven by software, there is flexibility for upgrades without changing hardware. In concept, the visual presentation on the monitor emulates that available from real equipment. Unless a functional mockup were to be used, however, the control configuration would not physically resemble the actual equipment or its controls. The absence of actual equipment and bridge configurations distinguishes desktop simulators from ship-bridge simulators.

Considerable advances have been made with respect to courseware (specially designed instructional software). Courseware design is either traditional show-and-tell for instructor-centered use, or interactive, with the student having a direct link to the software. Either form can include still graphics and the incorporation of embedded videos and simulations. Although application of these capabilities has been limited in marine transportation, there is a recent, rapid proliferation of microcomputer systems configured to support multimedia applications. The principal multimedia feature of such systems is the CD-ROM.

Hardware is quickly becoming a technological "nonissue." Interactive courseware capabilities in instructional systems can include branching subroutines that are keyed to student responses. Diagnostics can be embedded into the program to provide additional instruction, matched to the student's level of knowledge acquisition, to facilitate the learning process. These systems can be set up to accommodate the student's rate of learning. Interactive courseware has been developed for various applications within the U.S. Department of Defense (DOD) and commercially for use in mariner training.

Interactive classrooms can be used to improve student retention through instruction using interactive courseware. Interactive classrooms may be in any of several configurations. One form, which is instructor led, uses a button box or other simplified data-entry devices for student instruction and response. Another form includes a microcomputer workstation with a keyboard for each student. The instructor can electronically monitor student responses in either format.

Diagnostics can be included in the software to assist in determining whether the student has responded correctly or appropriately to lesson material or questions and how long it takes a student to get to the correct responses (in the case of tutorial-based lessons). The U.S. Naval Academy has established an interactive classroom environment (Bush, 1993). Similar instructional systems have not yet appeared for mariner training in general.

MICROCOMPUTER TRAINING ENVIRONMENT

In contrast to the operation of most ships, which is normally conducted while standing, a basic microcomputer desktop simulator requires that an individual sit at a workstation. The person participating in a desktop simulation is in a simplified training environment when compared to that aboard ship and in ship-bridge simulators. Placement of the trainee outside the normal bridge environment does not mean that desktop simulators are less effective for teaching and practice. One limitation of simulators, however, is their inability to successfully establish whether an individual can concurrently perform multiple tasks under the actual conditions found on a ship's bridge.

Interest is growing among marine research and manufacturing companies in the development of high-fidelity maneuvering simulations for automatic linkage into passage planning and execution. Several maneuvering simulations using bird's-eye views or simulated bridge window views, or a combination, are available. Computer software is also used to some extent as expert systems aboard a small number of commercial ships with integrated bridge systems to assist in decision making and to control special maneuvers, such as constant radius turns (NRC, 1994).

STANDARDS FOR SOFTWARE DEVELOPMENT

The development of microcomputer software for application in marine training and licensing is not guided by any industrywide technical performance or operational standards. Nevertheless, because of concerns over the possible misapplication of software instructional programs and the concerns over liability, developers generally appear to be taking a very careful approach.

Issues to be considered in the use of microcomputer simulators include:

- the possibility that desktop simulations might be treated as arcade games;
- the possibility that computer-assisted courseware might be used to "program" individuals to pass license examinations;

- the possibility that differences in the cue domain from actual conditions might create incorrect expectations about vessel maneuvering behavior or professional knowledge, skills, and proficiency; and
- the effectiveness of desktop simulators for evaluating human performance.

Development and use of standards might be adapted from other sectors. For example, DOD has established rigorous requirements to guide the development of interactive courseware to avoid redundancy and to conserve training development resources. Interactive courseware developed under contract to DOD must be recorded in a nationally accessible database. A search of this database is required as part of the development process for new interactive courseware.

RESEARCH, TECHNICAL, AND OPERATIONAL CONSIDERATIONS

General Research Basis

There are several research bases for using microcomputer simulators for marine training and licensing. These bases include general marine simulation research, U.S. Navy applied research in the development and use of interactive courseware and embedded simulations, and a small body of microcomputer-specific basic and applied research in the commercial marine transportation sector. Insights can also be obtained from human performance research in other sectors and potentially adapted to the marine setting (NRC, 1985; Hays and Singer, 1989).

The results of marine simulation research need to be carefully applied for several reasons:

- The research covers a period of great change and technological advances in both shipping and marine simulation.
- There is considerable variability in the simulators and methodologies used for research.
- The research basis consists of many experiments, a portion of which may not have been systematically confirmed by subsequent research, across a representative range of simulators, research methodologies, or through comparative analysis with actual operations.

Broad generalizations based on the existing research require considerable subjective interpretation. The veracity of such generalizations must be treated cautiously. Another important factor is that most of the published literature focuses on the use of ship-bridge rather than microcomputer desktop simulators. There is currently a very limited basis for comparing the relative merits across this range in simulator capabilities with respect to stimulation of mariner behavior in the same manner that mariners would perform operational functions and tasks in real life. Nevertheless, there is a substantial literature base on which to seek lessons that might be useful to the application of microcomputer simulations in professional development and marine licensing (Douwsma, 1993).

Computer-Aided Operations Research Facility Studies

Extensive mariner performance research was conducted by the U.S. Maritime Administration (MarAd) and the USCG from the mid-1970s through the mid-1980s using the ship-bridge simulator at the Computer Aided Operations Research Facility (CAORF), Kings Point, New York. This research initially focused on developing a clearer understanding of factors that affect human performance. The research methodologies that were employed were affected, to some extent, by the fact that researchers were learning how to use ship-bridge simulators in research. The simulator itself was sophisticated for its time. As a result, before the experiments could proceed, mariners who participated in the research had to learn how to accomplish certain critical tasks, such as measuring distances, in the simulator environment. Although the results of these experiments are a useful starting point, they have not been systematically updated with the changes in operating conditions in the merchant fleets.

Important issues that emerge from this research that need to be considered in the context of microcomputer simulators include:

- the adequacy of the visual scene and cues,
- the adequacy of the instrumentation cues,
- the relationship of cues to cognitive and motor skills,
- the accuracy and fidelity requirements, and
- user indoctrination to the simulator-based training environment.

Mariner Licensing Device

The USCG has been searching for ways to improve rules-of-the-road testing for a decade. The current multiple-choice, written examination format does not provide the applicant the opportunity to demonstrate the ability to interpret dynamic maneuvering information within the context of other navigation and bridge operational activities associated with the level of the license being sought. The rules-of-the-road test currently uses static graphic representations of maneuvering situations, lights, and shapes. The examinee is expected to analyze the situation and select and apply the appropriate rule from a short list of multiple-choice answers. The examinee is not required to detect and identify vessels or to determine changes in range and relative bearing while maintaining a complete perspective on the navigational and maneuvering situation. The provision to the examinee of range and bearing information and correct static interpretation of the data suggests, but does not verify, that the individual can effectively collect, interpret, and apply this information and integrate multiple sources of information during actual operations (ECO, 1987).

The USCG sponsored research and development of a prototype computer-based marine license testing device with embedded simulations. The goal was a device capable of authoring and administering examinations for:

- rules of the road;
- recognition of lights, shapes, and signals; and
- visual and radar navigation (ECO, 1987).

The testing objectives specified for the mariner licensing device are shown in Box G-1. Technical development specifications for the prototype are shown in Box G-2. The prototype system consisted of a modification of existing microcomputer ship-bridge simulators that had been previously developed for the U.S. Navy. The ship-bridge simulator featured a 90-degree video projection; ship

---

BOX G-1
Testing Objectives for Mariner License Testing Devices

- Applying appropriate rules of the road when in meeting, crossing, and overtaking situations under a variety of operational conditions.
- Applying the appropriate rules of the road when in special circumstances under a variety of operational conditions.
- Determining safe vessel speed under a variety of operational conditions.
- Shiphandling under various conditions of wind and current to:
  - hold course and speed to maintain a dead reckoning track,
  - avoid collision and pass at a safe distance with other traffic,
  - maneuver safely in various left and right turns within confined channels,
  - maneuver after a loss or degradation of propulsion power or steering in confined channels, and
  - to stop or slow.
- Shiphandling while executing emergency procedures.
- Using the whistle for maneuvering and warning signals under various operational situations.
- Using proper visual position-fixing techniques under various operational situations.
- Using proper radar navigation techniques under various operational situations.
- Recognizing and interpreting lights and shapes.
- Recognizing and interpreting sounds.
- Applying the appropriate rules of the road in restricted visibility.

SOURCE: ECO (1987).

> **BOX G-2**
> **Development Criteria for U.S. Coast Guard**
> **License Testing Devices**
>
> - Provision of hardware and software capable of authoring and administering mariner license examinations.
> - Visual, radar, and instrument display.
> - Input controls other than keyboards for executing rudder, engine, radar, and whistle signal actions.
> - Generation of visual and radar databases.
> - A computer-based instructional tutorial system for examinee use of the device.
> - "A flexible, inexpensive, and user-friendly examination authoring capability for continuous, dynamic simulation."
> - "A fairly 'non-controversial' automated scoring system."
> - A validation process involving scientific experiments with active mariners.
> - System technical and application documentation.
> - A comparative evaluation of the marine licensing testing device and the current licensing process.
>
> SOURCE: ECO (1987).

hydrodynamic mathematical models to drive the simulation; a console with a helm, throttles, and instruments; and a radar simulation. Hardware modifications to meet specifications for the prototype testing device included the addition of a visual bearing and "binocular-view" capabilities, whistle signal input controls, and a means for responding to multiple-choice questions (ECO, 1987).

The prototype was tested by senior licensed officers and operators of inland and coastal commercial vessels. Results were compared to written examinations. The test found that pass or fail results were affected by experience with computers and that prior experience on simulators did not prove significant. The research found no statistically significant differences in answers to questions with respect to any of the following: years of actual experience, level of education or current license, recency of experience, familiarity with computers, or experience with ship simulators. The mean test score for written examinations was about 90 percent, while the mean score for examination on the simulator was 20–25 percent lower, depending on the experience of the groups tested. About 70 percent of the mariners that participated in the testing program expressed a view that the simulation testing was superior to the current multiple-choice examinations in assessing actual capabilities (ECO, 1987).

An important result of the testing program was the indication that the ability of mariners to apply knowledge effectively, to maintain relatively complete situational awareness, to perform normal bridge functions, and to interpret all pertinent information in this larger context appears to be somewhat less than suggested by tests of knowledge alone. The gap between knowledge and the application of knowledge indicated that more realistic means of measuring ability to apply knowledge could improve the assessment of individual capabilities. Although the testing by itself did not improve the mariner's ability to close this gap, the success of the experiment suggests that the testing platform could be adapted to a training platform for this purpose. Test results also suggested that a more realistic appraisal of an applicant's competence with respect to applying rules-of-the-road knowledge was possible using a relatively compact simulation capability.

Interactive Video for Pilotage Training

The feasibility of using interactive video for pilotage training was jointly researched by the Cleveland Cliffs Iron Company and MarAd. The objective of the research was to determine whether such training could be substituted for some of the round trips over the Detroit, St. Clair, and St. Mary's rivers pilotage routes normally required for a licensed master or mate to receive a pilot's endorsement on a USCG-issued license. A major motivation for the research was that an individual on board to observe the route had to remain on board for its entire voyage from the upper to lower lakes, a minimum of four days, to observe one trip over the pilotage routes.

Extensive planning was conducted to support the production of videotapes of the pilotage routes and scripts. The videotaping covered various environmental conditions. A series of training tapes were made, each tape was about 60 minutes long and covered about 10 miles of pilotage route. Tapes were produced at the novice, intermediate, and mastery levels for each segment of the route. The novice level includes narrations, inserts on and highlights of key landmarks and aids to navigation, and descriptions of alternatives for making turns. Video overlay and taped questions requiring student response appeared at about 40-second intervals. Students were required to respond to a minimum number of questions correctly.

At the intermediate level, the inserts and highlights were omitted and the audio narration was more limited. The question format was more difficult. The number of allowable incorrect answers to questions was reduced. At the mastery level, there was no narration. All questions were in the form of video overlays. An incorrect answer terminated the session, and the participant was advised to repeat the novice session for that segment. Test subjects were cadets from the Great Lakes Maritime Academy (Townley et al., 1985).

The research was considered successful, although the small sample size and other factors limited the results to general observations. The results were considered a valuable starting point for proving the concept. It was determined (Townley et al., 1985) through the research that interactive video "appears to be an extremely poor substitute for other training aids in the development of chart sketching skills." It was determined, however, that "interactive video appears to be a cost-effective method of training pilotage candidates in preparation for shipboard observer time and as a means for reducing the trips necessary to master the waterway insofar as navigational (as versus shiphandling) skills are concerned." The interactive video format also appeared to be capable of imparting a fair degree of knowledge that was retained. The report recommended that further research be conducted to evaluate the potential of interactive video as a testing device for pilotage knowledge requirements. The report also identified and recommended a number of technical improvements for the production of interactive videos.

Maneuvering Simulations

Computer-generated imagery for microcomputer maneuvering simulations can take the form of simulated bridge window views or bird's-eye views on a maneuvering graphic or electronic chart, or a combination of both. Although development and use of a small-scale console is feasible, user instructions or reactions are normally entered via a keyboard, mouse, or trackball pointing device, rather than through an auxiliary console. Multiple monitors are feasible; however, single computer monitors are normally used to convey all information. The user usually must switch between views for visuals (up to four scenes—ahead, port, starboard, and astern), radar emulations, electronic charts, and instrument displays and controls (often displayed concurrently with other graphics). Switching between views places the visual scene in front of the trainee rather than requiring the trainee to physically move to obtain this view, as would occur during actual operations.

Vessel behavior is driven by a mathematical model that incorporates hydrodynamic reactions between the vessel and its operating environment. Environmental data are usually entered manually, although for shipboard applications automated input of some data may be feasible. The accuracy of microcomputer maneuvering simulations is directly dependent on the accuracy of trajectory predictions generated by the mathematical model and the bathymetric and environmental data used.

The general configuration and format of basic microcomputer desktop simulators is such that the mariner does not receive visual or physical cues in the same manner as when aboard ship. The degree to which these differences may affect performance—including interpretation, decision making, and leadership—has not been systematically researched. There is some indication, however, that a mariner's performance in a ship-bridge simulator may vary to some extent from performance during actual conditions. If this is so for a ship-bridge simulator that more closely replicates real-world conditions and cues, then it would be reasonable to suspect that similar effects would be associated with microcomputer maneuvering simulations.

Comparative analysis conducted by the Danish Maritime Institute (DMI) for real-life and simulated entries of a large passenger ferry to a new berth revealed that the swept path plots of simulated runs had closer tolerance than the real harbor entries. The report attributed this result to variable starting points for the real-life tests because of environmental conditions. The report also found that visual references were more frequently relied on in actual operations than in the simulator where there was more frequent use of the simulator's instrumentation. Based on responses to inquiries by the test's participants, this result was attributed to "problems with the estimation of speed and distance in the simulator." The heavier reliance on instruments in the simulation may have been responsible to some extent for the tighter tolerances, although examination of this possibility was not reported as a research objective.

Depth perception is an important consideration, because mariners tend to rely more heavily on electronic navigation equipment under conditions where distance cannot be reliably estimated from visual cues. More attention to electronic positioning systems may or may not be appropriate to improving marine safety. Simulation-induced reliance on electronic navigation equipment would, under the conditions of normal daylight operations in unrestricted visibility, mean that mariners were performing somewhat differently than they might when aboard a vessel in identical conditions. The significance, if any, between the actions that are evoked by the cue domain in a simulation and performance during actual operations is uncertain.

It is uncertain how human perception and cognitive responses might be affected by the microcomputer simulator format. The normal use of a single, small screen for representation of visual scenes and instrumentation and the limitations of graphics imagery in the single-monitor format make it reasonable to presume that users of microcomputer maneuvering simulations with simulated bridge window views would experience depth perception problems similar to or perhaps greater than those observed in the DMI experiment. Such problems could possibly be mitigated with radarlike bird's-eye views, especially where integrated with electronic charting systems. The cue domain would be altered from that associated with actual operations under unrestricted daylight operating conditions. The normal input devices, keyboards and pointing devices, are also different than those used aboard most ships. For training purposes, bridge team and bridge-to-bridge interactions are artificial because they are usually conveyed by text messages on the screen rather than by actual interactions among individuals. Because the cues are different from those experienced during actual operations, different cognitive skills or different levels of cognitive skills would be exercised.

Users of microcomputer simulators may find it necessary to rely more heavily on instrument emulations for position keeping and maneuvering. The resulting trajectories and swept paths would represent what can be achieved using a microcomputer simulation. How well the trajectories could be achieved during actual operations of all but the most sophisticated vessels on well-known routes is an open question. (It has been established through actual operations and extensive field testing involving passenger ferries serving ports in the Baltic Sea region that computers can be used to automatically maneuver ships on precise trajectories along well-known pilot routes [NRC, 1994]).

Because of the apparent necessity to rely heavily on instruments for maneuvering decisions and the tighter swept paths obtained using simulation during the DMI research, it is possible that better accuracy of maneuvering may be possible during desktop simulations than during actual operations. There are no data to determine whether this is the case; there are no data or research to determine whether the results of ship-bridge and desktop simulators are comparable; and there are no data to determine whether different cognitive skills are used to achieve the results. In the absence of full-bridge instrumentation, accurate replication of essential visual information, and well-defined job-task criteria to guide assessments, there is a very limited scientific basis for ascertaining which tasks or individual skills might be evaluated in a desktop simulator or whether or to what degree they could be correlated with actual operations.

The uncertainty over the results of desktop maneuvering simulations has implications for the application of this technology in passage planning. Although desktop simulators can potentially deliver accurate representations of maneuvering scenarios, there are uncertainties with respect to the degree to which the results represent vessel and mariner behavior in real life. The onboard maneuvering simulations would also be affected by the traffic conditions that exist at the time of passage, a factor that cannot be predicted for each individual transit.

REFERENCES

Anderson, D.B., T.L. Rice, R.G. Ross, J.D. Pendergraft, C.D. Kakuska, D.F. Meyers, S.J. Szczepaniak, and P.A. Stutman. 1993. Licensing 2000 and Beyond. Washington, D.C.: Office of Marine Safety, Security, and Environmental Protection, U.S. Coast Guard.

Bush, B. 1993. U.S. Naval Academy, personal communication, November 8.

Douwsma, D.G. 1993. Background Paper: Shiphandling Simulation Training. Unpublished literature review prepared for the Committee on Ship-Bridge Simulation Training, National Research Council, Washington, D.C.

ECO (Engineering Computer Optecnomics). 1987. Mariner Licensing Device. Final report. Contract No. DTCG23-86-C-30029. Washington, D.C.: U.S. Coast Guard.

Hays, R.T., and M.J. Singer. 1989. Simulation Fidelity in Training System Design: Bridging the Gap Between Reality and Training. New York: Springer-Verlag.

NRC (National Research Council). 1985. Human Factors Aspects of Simulation. E.R. Jones, R.T. Hennessy, and S. Deutsch, eds. Working Group on Simulation, Committee on Human Factors, Commission on Behavioral and Social Sciences and Education. Washington, D.C.: National Academy Press.

NRC (National Research Council). 1994. Minding the Helm: Marine Navigation and Piloting. Committee on Advances in Navigation and Piloting, Marine Board. Washington, D.C.: National Academy Press.

Townley, J.L., A. Wilson, and M.L. Thompson. 1985. Interactive Video Pilotage Training, Deck Officers: Final Report. Report No. MA-RD-770-85014. Washington, D.C.: U.S. Maritime Administration.

Erratum

Figures D-1, D-2, D-4 and Table D-1 were presented by H. Maeda and A.C. Landsburg at the Second International Symposium on Practical Design in Shipbuilding, convened by the Society of Naval Architects of Japan and Korea in 1983.